New Business Models and Value Creation: A Service Science Perspective

D1619904

Sxi – Springer per l'Innovazione

Sxi – Springer for Innovation

Volume 8

Sxi – Springer per l'Innovazione

Sxi – Springer for Innovation

http://www.springer.com/series/10062

Editor at Springer:
F. Bonadei
francesca.bonadei@springer.com

End of printing: November 2012

Lino Cinquini · Alberto Di Minin · Riccardo Varaldo
Editors

New Business Models and Value Creation: A Service Science Perspective

 Springer

Editors

Lino Cinquini
Istituto di Management
Scuola Superiore Sant'Anna
Pisa, Italy

Alberto Di Minin
Istituto di Management
Scuola Superiore Sant'Anna
Pisa, Italy

Riccardo Varaldo
Istituto di Management
Scuola Superiore Sant'Anna
Pisa, Italy

Sxi – Springer per l'Innovazione / Sxi – Springer for Innovation
ISSN 2239-2688   ISSN 2239-2696 (electronic)
ISBN 978-88-470-2837-1   ISBN 978-88-470-2838-8 (eBook)
DOI 10.1007/978-88-470-2838-8
Springer Milan Dordrecht Heidelberg London New York

Library of Congress Control Number: 2012950342

Translated and adapted from the Italian language edition: *Nuovi modelli di business e creazione di valore: la Scienza dei Servizi* by Lino Cinquini, Alberto Di Minin, Riccardo Varaldo 978-88-470-1844-0
© Springer-Verlag Italia 2011. All rights reserved

9 8 7 6 5 4 3 2 1

Cover design: Beatrice &, Milano

Typesetting: PTP-Berlin, Protago TeX-Production GmbH, Germany (www.ptp-berlin.eu)
Printing: Grafiche Porpora, Segrate (MI)

Printed in Italy

Springer-Verlag Italia S.r.l., Via Decembrio 28, I-20137 Milano
Springer is part of Springer Science+Business Media (www.springer.com)

Preface

The service-driven economy

In advanced economies, the modernisation and qualification of services are established factors that help increase the utilisation value of infrastructures and products, which expand and improve the overall supply of goods, acquiring competitive advantage through the stable integration of manufacture with innovative services. Due to their nature as "intrusive and pervasive production factors", services profusely participate in the growth of productivity and the overall competitiveness of the system and of industry itself. The challenge is therefore to enhance and promote the use of innovative services with the real industrialisation of the sector, as well as the forms of use and integration of services in different realities.

In recent years, in view of the extensive technological and organisational changes that have taken place in manufacturing, logistics and trade systems, academic and industrial research activities have developed to enhance the most powerful technologies and tools have been developed to help optimise services and their distribution in different businesses. Innovation functions to increase the levering effect in the modernisation of the economy – especially of industrial and logistics systems – by services. Their organic inclusion in supply chains involves frequently intervening and changing the architecture, governance and engineering of firm processes and compels innovation that combines technology, organisation and human resources as well as business and management models. In this sense, innovation in services is a powerful lever for cultural, technological and corporate renewal.

New models to create value

The theme of value and the strategies for its creation and sustainability is central in the economy of the firm and must be split into different dimensions: technical and managerial, measurement systems and business models. Creating value is the true goal of the economy: value for the firm, for the shareholder, for the customer and for the community. Without the ability to create value, no country can function and grow since increasing value is necessary for the production of wealth to be reinvested and redistributed.

The creation of value through services is a topic of great consequence. To create and capture value through services entails synergistically integrating and functionalising resources within and external to the firm's core business, as well as the country-system. This requires the ability to adopt advanced models of synergistic integration between firms and between the public and private sector, not only in financial terms, but also in terms of governance and management.

Under the pressure of globalisation, firms are required to review their organisational structures, business models and value chains, experimenting with more advanced ways of service utilisation. New value creation chains are inspired by the integration of manufacturing with services, enabling firms to be more efficient and more competitive. These chains open internationally according to a "diffused industry" model on a global basis, where emerging countries are destined to play an increasingly more meaningful role.

This new transnational economic and industrial strategy cannot be a medium to long-term strategy because the value generation model on which it rests is a project that requires time to obtain optimum use of resources through strategic partnerships and specific investments in different countries.

This is the new way of living and working in a globalised world. There are already signs, as well as significant cases, of firms moving in this direction: forward-thinking large and medium-sized firms launching ample and pervasive competitive repositioning, and implementing aggregations, mergers and integrations to achieve the new core-business dimensional thresholds imposed by globalisation. They are evolving the traditional model of growth and proprietary innovation by opening up to the outside; they are increasingly using open innovation that is progressively becoming an integral part of the business model for cutting-edge firms.

Innovation in services is also becoming more open, collaborative, multidisciplinary and global; the transformation of the business model becomes the turning point of this change and has an effect on value creation. This is the key challenge to exploit the opportunities of globalisation and international markets, making leaps forward in the innovation of processes, products and services, and influencing the nature and sustainability of competitive advantage.

Service science and the experience of the Scuola Superiore Sant'Anna of Pisa

Service science is the emerging discipline on an international level aiming to adapt and apply to the service sector scientific principles consolidated by management and industrial engineering while valorising its specificities as a distinguishing factor. In short, this is the interaction of different disciplines to form new professionals capable of designing, planning, organising and managing services in line with the requirements of an advanced and highly internationalised economy.

Service science provides an integrated and multidisciplinary approach to the amalgam of economics, management science, engineering and computer science to manage complex systems that are strongly impregnated with services in continuous adaptation and transformation when firms increasingly compete in the global economy. Within this framework, systemic collaboration between industries, universities and research is an obligatory step to enable and integrate the externalities of new knowledge and technology, and to form the new specialist competencies and managerial excellence needed by manufacturing and service industries.

A primary objective is to train people who are able to collaborate and interact with a large and internationalised innovation system, where valorising, making room for and giving voice to young talent is key, considering them as the fundamental drivers of value creation in the global economy. Furthermore, in this context, firms must be able to adopt advanced management systems, aimed at not only short-term objectives, but rendering it functional and appropriate to the strategy to be implemented. Management innovation is thus configured as a key component of the evolution of organisations and the valorisation of the human capital in the context of a globalised world.

It is the responsibility of universities to provide firms with human resources that are trained and ready to live and work in multicultural contexts, to assume a leadership role in organisational change driven by digitalised technologies and to develop new strategies for innovation and growth. This is particularly true in the Italian context, where the innovation culture in services is still marginal, affecting the belated modernisation of the entire sector (which accounts for around 80% of GDP). This is a serious limitation when considering that the development of advanced economies is now, more than ever, based on services.

In this respect, the Scuola Superiore Sant'Anna of Pisa in 2007 successfully launched a Master's of Innovation and Service Engineering (MAINS), putting to use twenty years of previous experience of training in innovation management, as a higher education programme to train people in developing resources of competencies and capabilities to operate successfully in service-oriented firms. More recently, in 2010, an International MSc in Innovation Management has also been established on the base of the MAINS knowledge assets.

In carrying out this project, a large group of leading companies is collaborating in the Sant'Anna School's Master's MAINS, with the aim of developing a systemic and original approach to service science that takes into account the living reality and is able to relate and integrate with it.

The contents of the chapters of this book, along with the cases developed in the MAINS "Innovation Lab" reported therein, fully confirm that it is possible to develop a systemic approach to service innovation in training, to design and implement new applications, and to improve organisational processes. All this may contribute to facing the innovation challenge by creating the cultural, professional and operational discontinuity needed to design new businesses and to launch a new economic development cycle.

Pisa, November 2012   Riccardo Varaldo
                                                   Agostino Sghedoni

Contents

List of contributors

Roberto Barontini (r.barontini@sssup.it) is Professor of Finance at the Scuola Superiore Sant'Anna of Pisa. His research activities at the Institute of Management include corporate governance and risk management, with articles published in national and international journals. He is Director of the MAINS Master's at the Scuola Superiore Sant'Anna.

Henry Chesbrough (chesbrou@haas.berkeley.edu) coined the term "open innovation" and has written the leading book *Open Innovation: The New Imperative for Creating and Profiting from Technology* (HBS Press, 2003). He is also the author of *Open: Business Models for Innovation*, recently translated into Italian (Egea, 2008). Chesbrough teaches at the Haas School of Business, University of California, Berkeley, where he is also Director of the Center for Open Innovation.

Lino Cinquini (l.cinquini@sssup.it) is Professor of Management Accounting and Business Administration at the Scuola Superiore Sant'Anna of Pisa. His research interests are cost management, strategic management accounting and performance measurement. He has conducted research in these areas in the manufacturing and service sectors. He is co-editor of the *Journal of Management and Governance*, Research Board Member of CIMA (UK) and co-director of the International Master's of Science MAIN (jointly run by the Scuola Superiore Sant'Anna and University of Trento).

Daniele Dalli (dalli@ec.unipi.it) teaches marketing at the Faculty of Economics, University of Pisa, and conducts research in consumer behaviour and consumer culture. He has published in national and international journals and is part of a European network (Coberen) dedicated to consumer studies.

Haluk Demirkan (haluk.demirkan@asu.edu) is Professor of Service Science & Information Systems and a research faculty member of the Center for Services Leadership at Arizona State University. His main research interests and expertise are in service science and innovation, service supply chain management and service-oriented sustainable solutions. In 2011, he was ranked 50th in the Top 100 Rankings of Worldwide Researchers according to the Association for Information Systems sanctioned research rankings.

Alberto Di Minin (a.diminin@sssup.it) is Assistant Professor of Management at the Scuola Superiore Sant'Anna and Research Fellow at the Berkeley Roundtable on the International Economy (BRIE, University of California, Berkeley). Alberto works at the School's Institute of Management and his research studies have been published in both national and international journals.

Marco Frey (m.frey@sssup.it) is Professor of Management at the Scuola Superiore Sant'Anna and conducts research activities on sustainability management issues, which have been published in national and international journals. He is Director of the Master's in Environmental Management and Control and Director of the Institute of Management of the Scuola Superiore Sant'Anna of Pisa.

Elie Geisler (geisler@stuart.iit.edu) is Distinguished Professor of Management at the Stuart School of Business, Illinois Institute of Technology (Chicago, USA), where she is Director of the Centre for Management of Medical Technology. She has authored 12 books and over 100 articles in international journals on the theme of technological innovation management, R&D evaluation, science and technology, knowledge management and management of biomedical technologies.

Riccardo Giannetti (rgiannet@ec.unipi.it) is Associate Professor of Management Accounting and Business Administration in the Faculty of Economics, University of Pisa. His research interests include cost and management accounting, on which he has published several articles and books.

Stephen K. Kwan (stephen.kwan@sjsu.edu) is Professor of Service Science and Management Information Systems, and Director of Online Programs Development in the College of Business at San Jose State University, USA. His current research interests include global trade in services, service systems and service value networks, and design thinking for service system innovation.

Riccardo Lanzara (rlanzara@ec.unipi.it) is Professor of Management at the Faculty of Economics, University of Pisa. He also teaches industrial marketing at the Faculty of Economics at the "Libera Università di Scienze Sociali" in Rome. His research interests include the evolution of production systems and the relationship between marketing and R&D. He has a leadership role in the Local Authority for Technology Transfer.

Giorgio Merli (g.merli@yahoo.it) is currently Professor of Business Sociology at the University of Milan Bicocca and Senior Vice President of Solving Ephesus, a global consultancy company. Previously managing partner of PricewaterhouseCoopers Consulting and IBM BCS, he is the author of numerous international books on innovative management issues.

Sabina Nuti (s.nuti@sssup.it) is Associate Professor of Management at the Scuola Superiore Sant'Anna of Pisa. She is the Director of the Management and Health

Research Laboratory (MeS) as a part of the Institute of Management. Her research activities include health management and performance evaluation. She has authored numerous publications in Italian and international journals.

Cinzia Panero (c.panero@sssup.it) is a researcher in management at the University of Genoa and collaborates in research projects with the MeS Laboratory of the Scuola Superiore Sant'Anna of Pisa on issues relating to health management.

Andrea Piccaluga (a.piccaluga@sssup.it) is Professor of Management at the Scuola Superiore Sant'Anna of Pisa where he coordinates the PhD in Management. Piccaluga is Scientific Director of the Scuola Internazionale di Alta Formazione of Volterra (SIAF), as well as part of the Institute of Management and is a member of the Board of Directors of the Italian Network for the Promotion of University Research (NetVal).

Guido M. Rey (g.rey@sssup.it) is Professor of Economics at the Scuola Superiore Sant'Anna of Pisa. He has taught at the Universities of Urbino-Ancona, Florence, Rome "La Sapienza" and Roma Tre. He entered the Bank of Italy as a bank official, later becoming a consultant, and was the President of ISTAT (National Statistic Institute) from 1980 to 1993 and President of AIPA (National Authority for Informatisation of the Public Sector) from 1993 to 2001.

Francesco Rizzi (f.rizzi@sssup.it) is a research fellow at the Scuola Superiore Sant'Anna and conducts research on energy and environmental management issues published in national and international journals.

Francesco Sandulli (sandulli@ccee.ucm.es) is Professor of Management at Università Complutense in Madrid. He carries out research activities on the impact of technology on the performance and structure of organisations as well as the relationship between innovation and strategy, which have been published in international journals. He has been visiting scholar at the Haas School of Business and Harvard Business School.

Agostino Sghedoni (a.sghedoni@sssup) is External Relations Advisor at the Scuola Superiore Sant'Anna of Pisa, collaborates in the organisation of its educational and research activities and is member of the MAINS Master's Steering Committee. Formerly marketing director of IBM Italy with responsibilities in marketing and management consulting and strategy, he is the author of several publications and interviews on the theme of enterprise transformation, competitiveness and internationalisation.

James C. Spohrer (spohrer@us.ibm.com) is IBM Innovation Champion and Director of IBM University Programs (IBM UP), and works to align IBM and universities globally for innovation amplification. Previously, Jim helped to found IBM's first Service Research group, the global Service Science community, and was founding CTO of IBM's Venture Capital Relations Group in Silicon Valley. During the 1990s, while at Apple Computer, he was awarded Apple's Distinguished Engineer Scientist and Technology title for his work on next generation learning platforms. His current research priorities include applying service science to study nested, networked holistic service systems, such as cities and universities. He has more than ninety publications and has been awarded nine patents.

Andrea Tenucci (a.tenucci@sssup.it) is a research fellow at the Scuola Superiore Sant'Anna. He conducts research activities at the Institute of Management on cost and management accounting and his works have been published in national and international journals.

Giuseppe Turchetti (g.turchetti@sssup.it) is Associate Professor of Management at the Scuola Superiore Sant'Anna of Pisa. He teaches economics and business management in the healthcare and insurance field, management of innovation in the biomedical field, and advanced marketing. His scientific interests include economics and management of service enterprises, innovation in the biomedical field and evaluation of healthcare services and technologies, topics on which he has published papers in national and international journals.

Riccardo Varaldo (varaldo@sssup.it) is Professor Emeritus of Management at the Scuola Superiore Sant'Anna of Pisa, in which he held the positions of Director and President. His main scientific interests are in economics and management of industrial enterprises, corporate strategy, internationalisation policies, economics and innovation management, SMEs and industrial districts. He has published numerous books and journal articles as well as presenting reports and participating in conferences. He is a member of the Institute of Management of the Scuola Superiore Sant'Anna, president of the Italian Society of Marketing and is part of the scientific committee of national and international journals.

Introduction

Lino Cinquini and Alberto Di Minin

This book constitutes an in-depth study of the theme of service science, the management area that the Scuola Superiore Sant'Anna – today organised within the newly born Institute of Management – reconfigured from the historic Innovation Master's launched in 1991 into the current Master's in Management, Innovation and Service Engineering (MAINS).

The interest in this research perspective has its foundations in the elements outlined in the Preface: the prevalence of the service sector in economic systems, the pervasive servitisation of economic and manufacturing activities, the innovation of traditional business model, and the development of new business models for value creation in the light of the new reconfiguration opportunities enabled by networks, ITC and many other enabling technologies.

The methodological approach and contents of service science are coherent with the focus on innovation in research and management education: innovation is today – and will increasingly be – an issue of service content and the ways of utilising it (Gallouj 2002). More and more companies look to services for new opportunities to differentiate themselves and avoid falling into the commodisation trap.

In this perspective, one aspect that renders the service science approach challenging to the complexity of social and economic organisations and their change is its multidisciplinary nature: in fact, the challenge of the complexity of organisation governance, both in terms of its technological–structural and dynamic–relational aspects, can be managed by developing specialist competency and integration capabilities as well as differentiated methodological approaches, where the technological and quantitative dimensions can merge with the capacity to understand the micro and macro phenomena on a qualitative level and the capability to orient their dynamics.

From this perspective, the development of new service economies once again indicates the centrality of the theme of human capital and its quality: service science helps understand and define the profile of competencies required by future managers and how it is important to invest in innovative training.

A brief analysis of the structure and content of this book will allow us to clarify this conceptual path.

The work is divided into two parts: the first defines some of the conceptual cornerstones of service science that constitute its foundation in the managerial perspective

L. Cinquini, A. Di Minin, R. Varaldo (eds.), *New Business Models and Value Creation: A Service Science Perspective*. Sxi 8, DOI 10.1007/978-88-470-2838-8, © Springer-Verlag Italia 2013

that is of interest to us. The second part reports some significant innovation experiences in service management models that are ascribable to this new approach.

Before browsing through the various contributions, we wish to indicate to the reader a part of the book strictly related to the educational experience of MAINS. In fact, a series of boxes have been inserted between the chapters reporting the experiences in actual cases subjected to in-depth study at the MAINS Master's Innovation Lab over the past four years.

The MAINS Laboratories are one of the most innovative aspects of the Master's curriculum and provide an output that transcends the didactic contribution to educational projects: organising experimental work in service economy and engineering through which students in multidisciplinary teams, which include both managers and teachers, learn to find solutions that create value.

The expected results may go beyond the "solution to a problem" and constitute a "contribution of ideas" to the development of innovative business models based on alliances that are often complex and challenging.

Therefore, the boxes scattered throughout the book constitute a unique wealth of experiences and ideas that enrich the contents, providing insights into the polyhedric nature of the application fields of service-oriented management approaches: from mobile payment to "servitisation", from multiservice cards to electronic invoicing, from 3.0 web-based organisations to the engineering of new products, from info-mobility to smart cities.

Turning to the contents of the chapters, the first chapter, "Service Science: On Reflection" (Spohrer, Kwan and Demirkan), provides, in the opening of the text, the most recent and updated framework of content as well as bibliographic references and key concepts that should guide reflection, constituting the definition of this discipline with a first attempt to integrate the different scientific areas that comprise it.

The articulation of the chapter, presented in this volume by three leading experts, gives the reader a sense of the complexity of service science and the ambitions of this multidisciplinary approach: as both practitioners and academics are converging in the effort to understand and govern the increasingly complex "service systems" in which economic and social activities are structured.

The conceptual foundations of service science addressed in the subsequent chapters of the first part of the book concern servitisation as a process that characterises the production processes applied to open innovation business models, the key role of users in the process of cocreating new value and the measurement issue in the contexts thus determined.

Dalli and Lanzara's contribution on "Product Servitisation" provides an updated framework of reference of the servitisation phenomenon, or rather, the gradual but significant increase of the service dimension in customer–supplier relationships, with particular reference to manufacturing firms and the implications of strategic positioning.

Chesbrough, Di Minin and Piccaluga, in their contribution "Business Model Innovation Paths", address the issue of innovation applied to business models, as important as technology today, and propose a logical framework that facilitates this business path. Merli's chapter, "The Transformation of the Business Model: Business Modelling", is also in line with the innovation and reconfiguration logic of business models by outsourcing decisions and/or network integration, through the presentation of a modular approach to analyse the strategic components that sustain innovation.

Sandulli's chapter, "User-Led Innovation: Final Users' Involvement in Value Cocreation in Services Industries", addresses the issue of open innovation starting from the firm's ability to activate its customers, which is of extreme strategic importance in service industries. It highlights the theme of cocreation of value that characterises the most advanced business models.

Finally, Barontini, Cinquini, Giannetti and Tenucci, in their chapter "Models of Performance and Value Measurement in Service Systems", indicate the main innovations that the called-for changes entail for models of measuring cost and performance, as well as the metrics that can better orient management towards the value creation objective.

The second part of the book is dedicated to exploring some innovative experiences in service management models that are particularly sensitive to these new strategic and management perspectives.

This part opens with a contribution by Rey on "ICTs as a Condition and as an Enabling Driver of Service Science in Italy". This is a preliminary reflection on the Italian case, considering the role of ICTs in developing conditions in the economic system for innovative models such as those subsequently proposed.

The health sector is the arena of the two subsequent contributions: "The Challenge of Healthcare Services: Between Process Standardisation and Service Customisation" by Nuti and Panero and "Home Healthcare Services: An Educative Case for the Development of a 'Service-Dominant Logic' Approach in the Marketing of High-Tech Services" by Turchetti and Gleiser.

The first of these two chapters explores the trade off between process standardisation and service customisation, a terrain particularly challenging in healthcare. Here, in fact, success depends on the competence to integrate two strategies that are usually alternatives in other sectors, through their explication in the concept of "appropriateness". The second contribution highlights the importance of appropriate marketing approaches that are able to grasp and overcome the problems, perhaps arising from the technology itself, which may hinder developing demand for a service; this issue is studied with respect to a homecare service, but takes on broader valence for service marketing.

Finally, Frey and Rizzi present the issues of innovation in service management models in the environmental and energy sector, where business models based on processes and service innovation are consolidating towards the supply chain in which the role of users is increasingly pivotal (e.g., in energy or water saving and recycling). At the same time, enterprises are consolidating approaches based on life cycle assessment, integrated management systems, and eco and energy efficiency.

In view of the structure of the contents, the question naturally arises as to whom the thoughts articulated in such a way are addressed.

We believe that various types of readers will be interested in these issues: business managers may discover a number of interesting indications, even at the operational level, thanks to the experiences reported in the chapters and also in the Innovation Lab cases reported; those involved in business development, control or marketing may find references and indications of interest in relation to their own activities and their evolution.

For policy makers too, some contributions in the text effectively highlight the characteristics that are profoundly changing organisations and their systems and how particular attention to services is essential in the political, economic and industrial infrastructure of country systems.

As for scholars and researchers, we hope that this work may contribute to a better understanding of the phenomena and trends that increasingly characterise the world of organisations in terms of the mode of functioning as well as the most effective way to govern and innovate.

In the research dimension, indications for the development of management research on these issues are just as numerous and significant: for example, the potential servitisation logic and its implications, Open Innovation and the related appropriability theme, business development in IT services, IT management control in open environments and the multidimensionality of performance. In this sense, the book may also constitute a research agenda with respect to the significant problems that human organisations will increasingly face in the near future.

As mentioned at the beginning, this book is the combined effort of research members and collaborators of the Institute of Management of the Scuola Superiore Sant'Anna of Pisa, involved in recent editions of the MAINS Master's and/or joint research on service science themes. The enthusiasm and commitment that as editors we have witnessed in bringing this publishing project to fruition testifies to the fact that this emerging area is considered extremely relevant for our research work, for future managers as well as students of Master's courses.

References

Gallouj F (2002) Innovation in the service economy: the new wealth of nations. Edward Elgar Publishing Ltd, Cheltenham

Part I
The fundamentals of service science

Service science: on reflection

1

James C. Spohrer, Stephen K. Kwan, and Haluk Demirkan

In this chapter, we reflect on the historical challenges and the future prospects for the scientific study of service phenomena. The growing dominance of the "service" component of national economies and corporate revenues has spurred expanded interest in service systems and service innovation. Service innovations based on global-scale information technology platforms are further fuelling interest in the scientific study of digitally connected service via cloud computing and smart phones. Practitioners and employers should be aware of and take advantage of the emergence of service science, as more and more universities around the world expand and deepen their service science-related curricula and research.

1.1
Introduction

For about ten years we have been working on creating a new transdiscipline to study service in business and society. A transdiscipline integrates aspects of many existing disciplines (e.g., service marketing, service operations, service computing, service economics, service design, etc.) without replacing any of them. We have seen the

J.C. Spohrer (✉)
Almaden Services Research, IBM Research – Almaden, San Jose, CA, USA
e-mail: spohrer@us.ibm.com

S.K. Kwan
College of Business, San Jose State University, San Jose, CA, USA
e-mail: stephen.kwan@sjsu.edu

H. Demirkan
The Center for Services Leadership, Arizona State University, Tempe, AZ, USA
e-mail: haluk.demirkan@asu.edu

L. Cinquini, A. Di Minin, R. Varaldo (eds.), *New Business Models and Value Creation: A Service Science Perspective.* Sxi 8, DOI 10.1007/978-88-470-2838-8_1, © Springer-Verlag Italia 2013

word "service" used in many different contexts – as adjective, noun and verb – and defined in as many different ways. Yet precisely defining and measuring real-world service phenomena for the purpose of establishing a "science of service" remains elusive. Of course, aspiring "service scientists" are in good company: after many decades biologists still cannot agree on their science's definition of life; cognitive scientists and computer scientists still cannot agree on their science's definition of intelligence; economists still cannot agree on their science's definition of value; and even though physicists routinely measure energy and work, they still debate definitions within their own domain. Intuitively, many service researchers sense a connection between a good definition of service and good definitions of quality of life, intelligence, value and work. Of course, the problems associated with precisely defining and measuring real-world phenomena should not be at all surprising; social scientists routinely remind scientists from other fields that the human enterprise of science itself evolves over time by shifting paradigms or agreed upon ways of perceiving and thinking about the world.

Nevertheless, service as the application of knowledge to create benefits for others naturally arises in the context of clearly identifiable entities, such as people, businesses and nations, that possess knowledge and have information-processing and communication capabilities as well as resource-based capabilities, including de facto and enforceable property rights. These diverse entities opportunistically and systematically interact to realise mutually beneficial outcomes (value cocreation). Simply put, service phenomena arise in a real-world ecology of entities, their interactions and their capacity for finding mutually beneficial outcomes.

All businesses apply knowledge for the benefit of others. All business entities are service businesses because all value is cocreated between interacting economic entities that possess information-processing and resource-integration capabilities put to work for the benefit of others (Vargo and Lusch 2004; Chesbrough and Spohrer 2006; Spohrer et al. 2007). Even modern manufacturing and agriculture can be seen as vast global supply and demand chains interconnecting entities. An economic entity (also known as a service system entity) is a dynamic configuration of resources, including people, technologies, organisations and shared information (Spohrer et al. 2007). Entities interact directly and indirectly by granting access rights to one another's resources (Spohrer and Maglio 2010).

For example, bus drivers transport passengers applying knowledge of driving and rules of the road; water utilities provide clean, safe drinking water to urbanites applying knowledge of purification and pressurisation; doctors tend to patients applying medical knowledge; and schools help students learn by applying staged curricular and pedagogical knowledge. These examples all illustrate entities engaged in routine service interactions. A sample list of what service can mean is depicted in Table 1.1.

In spite of the diversity, a focus on service framed in terms of entities, interactions and outcomes reveals some important commonalities. First, for complex service to evolve, service systems entities must have well defined rights and responsibilities to allow speedy dispute resolution. This allows entities to more readily experiment (e.g., succeed, fail, recover) and form extended networks as they interact and engage in service for service exchanges. Entities, such as people, businesses, universities,

Table 1.1 Different meanings of "service"

Service as a product
Hotels, telecommunication, IT, healthcare, financial, transportation
Customer service
Taking requests, answering customer questions, responding to complaints
Service derived from a tangible product
Autos (transportation service), mobile phones (communication service)
Service of product–dominant companies
Value added (e.g., repair and maintenance) to solutions (e.g., solving customer problems)

hospitals, etc., are partially contained within larger entities, such as cities, states and nations, that help enforce rights and responsibilities (Spohrer et al. 2007, 2011). They all apply knowledge, competencies and resources for the benefit of others, and engage in service for service exchange (value cocreation). For example, across nations, the populations can vary from hundreds of thousands (Iceland) to over a billion people (China), with differences in the structures associated with transportation, water, food, energy, communications, buildings, retail, finance, health, education and governance (Demirkan et al., 2011b, 2011c). In line with the service-dominant worldview (Vargo and Lusch 2004), the service system is viewed as the fundamental abstraction of the study of value cocreation or *service science* (IfM and IBM 2008; Maglio et al. 2010; Spohrer and Maglio 2010). The idea of non-zero-sum interactions that is value cocreation is not new (e.g., collaboration, cooperation, etc.): value emerges when entities work together for mutual benefit, the key being design or orchestration of these entities for effective value cocreation in constellations (e.g., Normann and Ramirez 1993) and networks (Gummesson 2004). The value that accrues is determined by entities' unique processes of valuing and derived in part from understanding the interactions of entities that are known as service systems. Service systems are physical symbol systems that compute the changing value of knowledge in the global service system ecology (Spohrer and Maglio 2008). Basically, all these entities can be viewed as service systems that depend on value-cocreation interactions to survive generation after generation and improve the quality of life of the people inside them by inventing better value-cocreation mechanisms for larger or smaller populations of entities in diverse environments. Typically they involve people, technology, organisations and shared information (also summarised as individuals, infrastructure, institutions and information). The types of shared information include such things as language, laws, measures, models and so on. They are connected internally and externally by value propositions, with governance mechanisms that support and adjudicate the resolution of disputes. Viability of entities within the ecology depends in part on their strategies for resource allocation and interaction with others, which influences their relative efficiency and growth or loss of knowledge-intensive capabilities (Spohrer and Maglio 2008).

In the interconnected world of today, if a nation, state or city were to become cut off from the rest of the world, quality of life for the population would begin to suffer almost immediately (e.g., punishment of governments and their populations through isolation and trade sanctions, embargoes). There is a much greater degree of interdependence among service system entities today than in the past. Nearly gone are the days when a pioneer family would set out with a wagonload of supplies and tools and fashion a self-sustaining family homestead in the wilderness. Instead, to-day's quality of life for most people (50% urbanites or city dwellers) is a function of the quality of service from many systems such as transportation, water, food, energy, communications, buildings, retail, finance, health, education and governance. In the world of today, quality of life is also a function of the quality of jobs (employment levels) in each of those systems. Furthermore, long term, quality of life is a function of the quality of investment opportunities available to improve those systems year over year, so each generation benefits from a rising standard of living. Nearly gone are the self-sufficient family-operated farms, mills, mines, fisheries or factories that use their local resources to create units of output to survive economically. Today, even something as simple as a modern pencil or toaster is beyond local family-operated production capabilities, and smartphones and spacecraft are beyond the capabilities of even most nations operating in isolation. Today, value is truly cocreated in networks of entities, best viewed as nested, networked service system entities for analysis and innovation purposes.

In the following sections, we trace the evolution of the concept of "service" in society, business and science from something to be aware of (and safely *ignored* in understanding the wealth of nations), to something to be measured (and *isolated* as different and often less controllable and less desirable for the idealised wealth of nations), to something to be innovated (and *integrated* back into our thinking about everything with respect to the sustainable wealth of nations), to being depended upon as the driver of growth (and put to the *forefront*). In conclusion, we present possible future research directions for service science. As an emerging transdiscipline, service science draws on many existing academic disciplines whose researchers work together to holistically integrate the separate disciplinary parts into a new whole, while enhancing, but not superseding, any of the parts.

1.2
Service being ignored

During the agricultural and early industrial ages, domestic and professional service activities were largely ignored by most but not all leading thinkers with theories on the wealth of nations. For example, the writings of five authors, Smith, Ricardo, Bastiat, Babbage, Ruskin and Chandler, provide a useful diverse sampling of conceptions of life, work and wealth of nations spanning the late 18th to the early 20th century. This time span corresponds to great transformations in average quality of life, as well as changes in typical institutions, technologies and skills in the USA and Europe. These five authors wrote about the way nations created wealth based on human strengths and limitations – physical, mental and social (or moral) effort.

Smith (1776/1904) advocated increased specialisation. Smith argued that efficiently employing growing populations in the production of material goods was essential to the wealth of nations. Through division of labour, illustrated by the manufacture of pins, productive labour exhibits increasing returns (e.g., specialised labour has higher skill levels, less time lost in shifting occupations and greater opportunity for discoveries). In contrast, musical and entertainment performances as well as several other human-oriented service activities (including churchmen and men of letters) illustrated unproductive labour that could not be profitably scaled in the same way to increase units of output in a given time.

Ricardo (1817/2004) extended Smith's notion of division of labour between people to trade between nations. Ricardo clarified the concept of comparative advantage, and the opportunity to better balance the relative activity levels across the spectrum of national productive activities and investments. However, the focus was primarily on units of agricultural and tchnological imports and exports, and not on immigration, flows of people or flows of information between nations.

Bastiat (1850/1979) criticised the views of Smith and Ricardo on value and wealth of nations, claiming that even simple material things embody the accumulated knowledge of many people. Bastiat was one of the first to recognise that all exchanges can be viewed as direct or indirect human knowledge-based service for service exchanges, even when material things or money are exchanged, because in reality things embody the present or past applied knowledge and effort of people working to benefit themselves and others, i.e., service.

Babbage (1832/2012) wrote about the economics of machinery "to supersede the skill and power of the human arm". Babbage advocated that the wealthiest in society might benefit from a deeper understanding of the machinery on which their wealth and leisure was based, and then better invest to realise future prospects of even more advanced machinery. Babbage also studied the effect of machinery on reducing demand for labour; especially specialised manufacturing labour of the type Smith's prescriptions would create. Babbage did consider the high-skill knowledge-based labour of engineers and scientists that created machinery to replace human physical and cognitive labour to be a prized type of professional service for the wealth of nations.

More in line with Bastiat regarding the importance of service, Ruskin (1860/2012) concluded that the ultimate wealth of nations was its people. Ruskin wrote about the five great professions of civilised nations from the perspective of their citizens: soldiers defend, pastors teach, physicians heal, lawyers enforce justice and merchants provide. If free to immigrate, people, especially those disenfranchised in society, those who suffer or are oppressed, would move to civilised nations where the best of these service professionals practised. Ruskin suggested great professionals are defined by the great challenges they willingly take on in their work, including when to die in service of others: "For, truly, the man who does not know when to die, does not know how to live". Gandhi was greatly influenced by Ruskin's work, which he found "impossible to lay aside".

Chandler (1977) documented the rise of professional managers and the service they provided to large firms (institutions) as a source of increased wealth for nations. The hierarchical structure of a nation's military, churches and judiciary provided a model for transportation, communication, energy, finance and other commercial industries at a national and global scale.

Within Smith's cultural context, the view that many service activities were examples of "unproductive labour" and therefore defied profitable scaling by division of labour seemed like common sense. At the time, a high quality of life from material goods was enjoyed by only a few who were in positions to directly or indirectly control the labour of others. High quality of life through material wealth was reserved for relatively few, who owned substantial property through inheritance; or owned productive land with servants, labourers or large families to work it; or possessed the skills and tools of a successful craftsperson; or were shopkeepers ideally near a town or city; or had the social contacts, skills and ambition to associate with scale-up businesses of the times, mostly in manufacturing or distribution. Many people were poorly educated and worked physically exhausting, hard lives, and many died relatively young from ailments easily cured today. Cities offered expanded opportunities, but unhealthy conditions spread disease and sickness, especially among the poor. The average person could either stay rooted to their home community and family, or seek out opportunities with associated higher levels of risk in cities, the military or the seas. In this era, the unimportance of service relegated it to the bottom of the society.

1.3
Service being isolated

As the 20th century unfolded, Clark, Baumol, Leavitt, Zysman and Sen provided a set of updated perspectives on the changing nature of life, work and wealth of nations. Through their writings and those of other influencers, service activities in business and society were becoming a legitimate, economically and socially significant area of study to be recognised and isolated from other activities and subject to empirical studies.

Clark (1940/1957) challenged the status quo of largely ignoring the "service sector" in assessing the wealth of nations. In nations with growing wealth and quality of life for average citizens, government (defending and enforcing justice), education (teaching), health (healing), retail (providing), financial, transportation, communication, public utility and even entertainment service activities significantly increased, as anticipated by Ruskin. With increased automation of agriculture and manufacturing, human labour was shifting to professional service activities, often employing science and engineering skills in support of agriculture and manufacturing, as anticipated by Babbage. With growing wealth and concomitant desire for more leisure time, families were able to afford to outsource more and more domestic service activities, as anticipated by Bastiat.

Baumol and Bowen (1966) sounded the alarm first issued by Smith: if the wealth of a nation depended on two sectors, one in which labour productivity increased over time (such as manufacturing activities) and one in which labour productivity was stagnant (such as string quartets and other service activities), over time all labour would be absorbed into the stagnant sector and the growth of national wealth would slow. However, by this time, in the mid-20th century, technology for recording and distributing music was entering the technology infrastructure of nations, even homes and automobiles. Baumol (2002), recognising the increase in complexity of the environment, later revised his model, as anticipated by Babbage, to include the effects of research on improving the labour productivity even in supposedly stagnant sectors. Perhaps ignoring or working to slow the growth of the service sector was not the best policy after all. Perhaps it was better to isolate it and study it separately, developing new methods of improving the efficiency of service activities.

Levitt (1976) echoed the need to study efficiency improvements in service activities. Leavitt used fast-food restaurants as an example of new levels of efficiency and labour productivity coming to service businesses. Furthermore, Leavitt challenged the notion that a business could be classified simply as a manufacturing business (e.g., IBM) or a service business (e.g., Citibank), suggesting that all businesses had some degree of front-office customer-facing service activities and some degree of back-office internal activities more amenable to efficiency improvements. Levitt (1972) asserted that "There are no such things as service industries. There are only industries whose service components are greater or lesser than those of other industries. Everybody is in service."

Cohen and Zysman (1988) extended the ideas of Leavitt beyond the boundaries of a single firm, noting that manufacturing activities and service activities were connected in upstream and downstream networks. Specialisation was blurring distinctions and creating longer and more intricate networks of service and manufacturing businesses. Zysman further suggested that outsourcing service activities might be more difficult where customer contact was required, and that reducing certain manufacturing activities in the value network to benefit from Ricardo's law of comparative advantage might have a deskilling effect on nations that could inhibit future innovations and negatively impact the wealth of nations.

Sen (2000) greatly expanded on a theme in Ruskin that increasing the capabilities of people, including their freedoms and opportunities, was not only a moral imperative, but essential to the wealth of nations. Sen investigated in empirical studies the interconnections between (1) political freedoms (enforce justice), (2) economic facilities (provide), (3) social opportunities (heal and teach), (4) transparency guarantees (enforce justice) and (5) protective security (defend). Again the logic of people voting with their feet is practical evidence for this position.

By the end of the 20th century, service was no longer being ignored; it had been rediscovered, purified and isolated by many as an important component of the wealth of nations. In fact, beyond policymakers and economists, many other academic disciplines had begun to zero in on service activities as an understudied area. Marketing, operations management, operations research, industrial and systems engineering, management of information systems, computer science, statistics, management of innovation, and design were all working to ensure their graduates hired into service businesses or product businesses (with service offerings, such as finance, maintenance, field service, etc.) were better prepared. The quality of life of the average citizen, skills levels, types of institutions and the technology infrastructure available to citizens in wealthy nations were greatly transformed. As Sen commented: "We live in a world of unprecedented opulence, of a kind that would be hard even to imagine a century or two ago". The notions put forward by the aforementioned authors were made more complex later on in the century and into the 21st century when transborder trade in services became a more heated battle ground than conflicts over the trade of manufactured goods and farm products.

1.4
Service being integrated

As the 21st century unfolds, governments, universities and companies have begun to integrate the work from separate disciplines into a new area called service science with a focus on service innovation (IfM and IBM 2008; UK Royal Society 2009; Ostrom et al. 2010; IBM 2011; IBM Research 2004).

Service innovations can improve customer–provider interactions and outcomes as well as stakeholder opportunities for colearning and codeveloping. Service innovations increase both domain-specific knowledge and the ability of stakeholders to apply that knowledge to realise mutual benefits. The need for both new domain knowledge and application knowledge is often summarised as "invention does not equal innovation". Inventions are potential pathways for value cocreation (the "how to recipes" that are novel, non-obvious and useful), whereas great innovations must create substantial benefits in business or society (the "actual feasts" that sustainably satisfy needs, wants or aspirations for some period of time). Service innovations are new and improved service inventions put into practice and scaled up rapidly; all this requires a deeper understanding of service systems. Service science aims to clarify "what is service innovation?"

When first approaching the question, one finds few very relevant definitions of service innovation. The Finnish Funding Agency for Technology and Innovation defines service innovation as "…a new or significantly improved service concept that is taken into practice. It can be for example a new customer interaction channel, a distribution system or a technological concept or a combination of them. A service innovation benefits both the service producer and customers and it improves its developer's competitive edge." Also, Ostrom et al. (2010) defines it as "…a service innovation can be viewed by some as a cost-efficient way to streamline information exchanges, reduce mistakes, and ensure targeted levels of service quality. Others perceive it as a loss of responsiveness and personal discretion that endangers job security and is detrimental to employee motivation and customer satisfaction."

Basically, service innovation is based on evolving customer behaviours and market trends. These changes are hard to predict accurately and the success of a service can hinge upon a small nuance that is hard to pinpoint (Demirkan 2010).

As a discipline, service science starts with the practical study of stakeholders and their interaction patterns and outcomes. Stakeholder entities' (e.g., bus drivers/passengers, water utilities/urbanites, doctors/patients, schools/students, etc.) interaction patterns involve systematic resource access (as part of transportation, utilities, health, education systems, etc.), which generate outcomes, both mutual benefits and non-benefits that arise as either sustainable or unsustainable value-cocreation phenomena. Service science studies the evolution of the ecology of service systems, made up of stakeholder entities, interactions and outcomes, over multiple generations (Demirkan et al. 2011a). A key part of the historical high-level view of service science is its multiple generation perspective.

The most noteworthy societal service innovations in history, such as good governance, healthcare and education, both increase sustainable value-cocreation between entities and improve average quality of life for individuals, generation over generation. For our purposes, formal service system entities (e.g., people, businesses and nations) are recognised as legal entities by governments, have legal rights and responsibilities to make, keep and break contracts, and are characterised by bounded rationality and path dependence with respect to their symbolic reasoning, communication and learning capabilities. Informal service systems may or may not be recognised as legal entities by governments, but are composed of people with capabilities to access and exchange things and ideas (resources and more), as well as make, keep and break promises using natural language skills.

Many members of the emerging service science community agree that one of the most fundamental and sustainable service innovation (thus value cocreation) mechanisms studied so far combines Ricardo's law of association (comparative advantage) with human learning from experience (learning curves). Understanding this value-cocreation mechanism is not intuitive to many and requires study.

For example, and paradoxically, if one "superior entity" does all activities better than other entities (uses less time and resources to achieve all goals), in most cases as long as there are variations in relative performance capabilities across activities/goals, it makes mathematical sense for that superior entity to interact with others to reap the benefit of the others' inferior application of knowledge, or service capabilities! Through appropriate interactions, which heuristically allow each entity to do a little more of what it does best and a little less of what it does least well, all entities can use interactions and exchanges to realise both a small individual and large collective gains (saving time and resources).

Doing more of what one "does best" also accelerates learning from experience (more experience, more improvement) and therefore the individual and collective benefits get bigger rapidly over time, up to a limit, as entities become more specialised and dependent on complex technologies and rules of interaction to function well collectively. The limit is relatively quickly reached when overspecialisation and reliance on complex technologies and rules creates a "knowledge burden" for entities with finite lifespans that must transfer knowledge from one generation to the next generation of entities, and when overspecialisation in large networks of inter-dependent entities lacks resiliency to the loss of individuals and groups of entities. Ecological collapse from loss of a single species is a biological equivalent. Most people think of ecology in terms of living organisms, like plants and animals in a natural environment. However, the concept of ecology is more general and can be applied to entities as diverse as the populations of types of atoms in stars to the types of businesses in a national economy. Ecologies of entities and their interactions are, in general, complex objects of study on par with the human-made service system entities and human-made value-cocreation mechanisms.

The learning and exchange of new technologies (infrastructure) and new systems of rules (institutions) can have a profound effect on the knowledge people (individuals) possess and apply to achieve mutual benefits within a changing cultural context of what prosperity and quality of life mean individually and collectively (shared information). Service science studies the ever evolving ecology of service system entities, which (1) act in diverse stakeholder roles – customers, providers, authorities and competitors; (2) dynamically reconfigure access to diverse resources – people (individuals with skills and expectations), technology (infrastructure), organisations (institutions with roles and rules) and cultural context (shared information); (3) have value proposition-based and governance mechanism-based interactions; and (4) transform expected and realised value-cocreation outcomes in path-dependent ways.

Spohrer and Kwan (2009) define innovation as a measure of value (judgement of change). They also indicate that "innovation in service system ecologies (multiple populations of interacting service systems) is a relative measure of the value co-creation increase that results from a change. The types of changes can be the creation of new types or instances of resources, service systems, value propositions or governance mechanisms. The types of measures of value in rationally designed service system improvements relate to the four fundamental measures of quality (customers), productivity (providers), regulatory compliance (authorities), and sustainable innovation (competitors). Examples of service system innovation include: (1) a loyalty program for an airline, (2) a self-service system at a bank (ATMs), airport (tickets), or retail outlet (checkout scanning), (3) creating a financial services offering, (4) creating a new franchise model, (5) creating a new type of business or organisational structure, (6) specialising and streamlining a medical procedure to expand the number of patients that can afford and hence seek treatment, etc."

Moreover, service science is an emerging transdiscipline that borrows from and tries to re-integrate multiple disciplines, such as marketing and behavioural sciences (customers), operations research and management sciences (providers), governance and political science (authorities), game theory and learning sciences (competitors), psychology and cognitive science (people), industrial engineering and system sciences (technology), management of information systems and computer science (information), organisation theory and administrative sciences (organisations), economics, law and historical sciences (past changes), foresight studies and design science (future possible changes), time use and labour studies, in light of anthropology and demographic sciences (practice of what is presently valued most and how that shifts through the ages in different societies and contexts), to name a few of the many disciplines that service science draws on (but does not supersede) (Demirkan and Spohrer 2010).

1.5
Service being put to the forefront

Today, many modern service offerings are information and communication technology (ICT)-enabled. The convergence of ICT innovation, design, development, execution, storage, transmission and reusable knowledge is creating new opportunities (Demirkan et al. 2009). They include redeploying people, reconfiguring organisations, sharing information (e.g., language, processes, metrics, prices, policies and laws) and investing in technologies. More specifically, ICT provides the means to improve the efficiency, effectiveness and innovativeness of organisations through: (1) making it possible for commoditisation of none-core competencies (e.g., outsourcing, out-tasking); (2) improving collaboration (e.g., inter- and intra-organisational workflows and business processes); (3) decreasing the risk of information security breaches; (4) facilitation of new types of services (e.g., Google, online banking); (5) separation of production and consumption of a service, thus storability, transportability and access to knowledge-based services (e.g., tax software, online classes); (6) coordination of service systems (e.g., online broker systems, information markets, open innovation platforms); (7) reduction of the costs of service production (e.g., semi- and fully automated call centres); (8) improvement of customer-perceived service quality (e.g., ability to standardise elements of service as well as customise to the individual when appropriate); and (9) integration of customers into service creation and delivery (e.g., online educational services, health information systems, business-to-business solutions) (Allen et al. 2006; Bræk and Floch 2005; Davenport 2005; Garcia-Murillo and MacInnes 2003; Garrison, 2000; Maglio et al. 2006; Soper et al. 2007). As the world becomes digitally connected, the significance of customer cocreation, open innovation and service-scaling phenomena is becoming more apparent, and empirical investigations can draw on new data sets and tools (Bardhan et al. 2010; Hsu and Spohrer 2009; Chesbrough 2011). ICT innovations are making technology more service oriented and also enabling many service innovations.

Basically business-to-consumer (B2C), business-to-business (B2B), business-to-government (B2G) and business-to-business-to-customer (B2B2C) service offerings, and many versions of value-cocreation configurations have been used more frequently. All services – from knowledge-intensive professions (e.g., business consultant, physician, software engineer, legal counsel, financial advisor, teacher) to information systems and technology services (e.g., self-service technologies, business services, web services, software-as-a service, infrastructure services, virtual computer resources, grid computing, cloud computing) – have characteristics that scholars have studied, looking for salient patterns (Bitner and Brown 2006; Demirkan et al. 2009; Fitzsimmons and Fitzsimmons 2007; Looy et al. 1998).

- Many service offerings and interactions include high involvement of people for delivery and usage. Human are the primary resources and stakeholders. While people can be unpredictable in their behaviour, the planning, design, delivery and support of any service require variability, heterogeneity or non-standardisation.
- Service offerings and interactions are mostly something one cannot touch or feel, although they may be associated with something physical. For example, while receiving an education service, we also use books, notes, etc.
- Unlike products, service offerings and interactions cannot be stored in inventory for later use. Therefore, management of demand and capacity, and pricing decisions are very crucial in the provision of services.
- Many service offerings and interactions are produced and consumed simultaneously. This will also result in role interactions rather than things.
- Many service offerings and interactions are delivered to customers with the collaboration of distributed service providers.
- Simultaneity of production and consumption of service offerings and interactions occurs in complex service environments due to interaction of people, processes, technology and shared information.

Today's highly competitive, global economy, with high customer expectations, dynamically changing markets and technologies, and a fragmented regulatory environment, increases the need for companies to be able to de-commoditise their assets and have transitioned from a focus on goods to a focus on service offerings and interactions (Christensen and Raynor 2003; Vargo and Lusch 2004). The implications of this shift in offerings and interactions to technology innovation and management practice are dramatic (Table 1.2).

Table 1.2 Transition from past to present (revised from Demirkan et al. 2009)

From	To
Focus on goods	Focus on service
Cost reduction through manufacturing efficiency	Revenue expansion through service
Standardisation	Customisation
Mass marketing	One-on-one marketing
Short-term transactions	Long-term relationships
Function oriented	Coordination oriented
Limited information-sharing capabilities	Improved information-sharing capabilities
Application silos	Integrated solutions
Tightly coupled businesses	Loosely coupled solutions
Sale-contracts	Rental-service level agreements

1.6
Concluding remarks: future directions

The growth of the service sector of the economy is truly a wonder of human history, on par with the agricultural revolution and the industrial revolution. Practitioners with an interest in service innovation should be aware of service science, and monitor the advancements in service-science related curricula and research results. The field is rapidly evolving. The numbers of "service"-focused schools, programmes, classes, and academic and professional associations, articles, conferences, talks and news are growing exponentially.

Nevertheless, after just ten years, service science is still in its infancy. Typically, new scientific disciplines take about 30 years to mature. For example, computer science is now a mature discipline, but when it started in the mid-1940s there were no semiconductors, transistors, integrated circuits, software development tools or authoritative compilations of algorithms, and no Internet. Until universities find a faster way to create new PhD students, it will probably always take about 30 years, which is three to five generations of students, to create a new discipline, especially a complex transdiscipline like service science.

As we look to the future, three advances are on the horizon: (1) new tools and design specification standards, (2) new open data sources and streams, and (3) new on-line textbooks with dynamic real-world problem sets.

First, tools such as computer-aided design (CAD) systems for modelling and designing complex service systems are being developed. Nearly every product manufactured today is designed with a CAD system and has an associated CAD file. However, service offerings rarely have standard specifications. From enterprise architecture to business process specifications to new tools for designing buildings and whole city simulations, this area of tool development and standard specification of service systems and processes is poised for change. These tools may be especially valuable to policymakers working to understand regional innovation systems, especially if alternative policies can be explored (Spohrer and Giuiusa 2012).

Second, new open data sources and streams are becoming available. Businesses and governments are working to make more and more data sets available to researchers. For example, the new company Kaggle (http://www.kaggle.com) invites researchers with the slogan "We make data science a sport". "Big data" initiatives in many nations are providing the data sets for complex service systems and stakeholder interactions to perform more and better empirical investigations, faster and cheaper (Demirkan and Delen 2012). Big data initiatives are growing because of advances made in the collecting, processing and analysis of lots of data.

Third, while textbooks may seem like a 20th-century concept, disciplines do advance by the creation of professional associates, and "books of knowledge" that distil the concepts and best cases to help students rapidly learn about new fields. More and more relevant textbooks and teaching materials for learning aspects of service science are appearing ever year (Spohrer 2012). Also, re-invented on-line textbooks with dynamic problem sets that use on-line data sources and CAD tools will link these three advances, expected in the coming decade.

The goal of service science is to catalogue and understand service systems, and to apply that understanding to advancing our ability to design, improve and scale service systems for practical business and societal purposes (Demirkan et al. 2009). The growth of service economies has broad implications for the operation of businesses, the creation of academic knowledge, the delivery of education, the implementation of government policies and the pursuit of humanitarian causes.

In sum, the measured growth of "service" in society and business parallels a growing global dependence on smarter (knowledge-intensive, digital) systems. Smarter systems integrate and apply knowledge (people and skills, technology and business models) to realise benefits for customers, and so they can be referred to as product service systems or simply "service systems". Furthermore, the growth of complex nested, networked service systems also parallels trends in service outsourcing, self-service technologies, globalisation, urbanisation, technological augmentation of human labour, knowledge intensity of high-skill work, and rewards for entrepreneurial capitalism, especially when business models allow innovations to scale rapidly. Service science aims to provide fundamental knowledge and tools to better understand service innovation scale-out and thereby shape public policy to guide the evolution of interconnected service systems that improve average quality-of-life globally, generation after generation. Service science has come far after ten years. It is hoped that by working together better it will not take another twenty years for service science to become mature and make a positive and sustainable impact on society.

References

Allen P, Higgins S, McRaie P, Schlaman H (2006) Service orientation: winning strategy and best practices. Cambridge University Press, New York

Babbage C (1832/2012) On the economy of machinery and manufactures. Amazon Digital Services. A Public Domain Book. Kindle Edition

Bardhan I, Demirkan H, Kannan PK et al (2010) An interdisciplinary perspective on IT services management and services science. J Manag Inf Syst 26(4):13–65

Bastiat F (1850/1979) Economic harmonies. The Foundation for Economics Education, Irvington-on-Hudson

Baumol, WJ (2002) Services as leaders and the leader of the services. In: Gadrey J, Gallouj F (eds) Productivity, innovation and knowledge in services: new economic and socio-economic approaches. Edward Elgar, Cheltenham

Baumol WJ, Bowen WG (1966) Performing arts: the economic dilemma. The Twentieth Century Fund, New York

Bitner MJ, Brown SW (2006) The evolution and discovery of services science in business schools. Commun ACM 49(7):73–78

Bræk R, Floch J (2005) ICT convergence: modeling issues. In: Amyot D, Williams AW (eds) Proceedings of the 4th International SDL and MSC Workshop, Ottawa, Canada, June 2004, Lecture Notes in Computer Science, 3319. Springer, Berlin

Chandler AD (1977) The visible hand: the managerial revolution in American business. Belknap/Harvard University Press, Cambridge, MA

Chesbrough H (2011) Open services innovation: rethinking your business to grow and compete in the new era, 1st edn. Jossey-Bass, Wiley, San Francisco

Chesbrough H, Spohrer J (2006) A research manifesto for services science. Commun ACM 49(7):35–40

Christensen CM, Raynor ME (2003) How to avoid commoditization. Harvard Business School Press, Boston

Clark C (1940/1957) Conditions of economic progress, 3rd edn. Macmillan, New York

Cohen SS, Zysman J (1988) Manufacturing matters: the myth of the post-industrial economy. Basic Books, New York

Davenport TH (2005) The coming commoditization of processes. Harvard Bus Rev 83(6):100–108

Demirkan H (2010) Service innovation and the science of service. Korea Institute for Industrial Economics & Trade (KIET) Innovations in Service Industries Conference, Seoul, Korea, November 16

Demirkan H, Delen D (2012) Leveraging the capabilities of service-oriented decision support systems: putting analytics and big data in cloud. Decis Support Syst Electron Commer (In press)

Demirkan H, Spohrer JC (2010) Servitized enterprises for distributed collaborative commerce. Int J Serv Sci Man Eng Technol 1(1):68–81

Demirkan H, Kauffman RJ, Vayghan JA et al. (2009) Service-oriented technology and management: perspectives on research and practice for the coming decade. Electron Commer Res Appl J 7:356–376

Demirkan H, Spohrer JC, Krishna V (2011a) Service and science. In: Demirkan H, Spohrer JC, Krishna V (eds), The science of service systems (Series: Service science: research and innovations in the service economy). Springer Science+Business Media

Demirkan H, Spohrer J, Krishna V (eds) (2011b) The science of service systems (Series: Service science: research and innovations in the service economy). Springer Science+Business Media

Demirkan H, Spohrer J, Krishna V (eds) (2011c) Service systems implementation (Series: Service science: research and innovations in the service economy). Springer Science+Business Media

Fitzsimmons JA, Fitzsimmons MJ (2007) Service management: operations, strategy, information technology, 6th edn. McGraw-Hill-Irwin, New York

Garcia-Murillo M, MacInnes I (2003) The impact of technological convergence on the regulation of ICT industries. Int J Media Manag 5:57–67

Garrison B (2000) Convergence to the Web is no longer just the future. College Media Rev 38(3):28–30

Gummesson E (2004) Return on relationships (ROR): the value of relationship marketing and CRM in business-to-business contexts. J Bus Ind Mark 19(2):136

Hsu C, Spohrer JC (2009) Improving service quality and productivity: exploring the digital connections scaling model. Int J Serv Technol Manag 11(3):272–292

IBM (2011) The invention of service science. IBM Centennial Celebration, IBM 100 Icons of Progress. Available at http://www-03.ibm.com/ibm/history/ibm100/us/en/icons/service science. Last accessed on September 13, 2012

IBM Research (2004) Services science: a new academic discipline? The architecture of on-demand. Business Summit, T.J. Watson Research Center, IBM Corporation, Yorktown Heights, NY

IfM and IBM (2008) Succeeding through service innovation: a service perspective for education, research, business and government. University of Cambridge Institute for Manufacturing, Cambridge

Levitt T (1972) Production-line approach to service. Harvard Bus Rev September/October

Levitt T (1976) The industrialization of service. Harvard Bus Rev 54(5):63–74

Looy BV, Dierdonck RV, Gemmel P (1998) Services management: an integrated approach. Financial Times Pitman Publishing, London

Maglio PP, Srinivasan S, Kreulen JT, Spohrer J (2006) Service systems, service scientists, SSME, and innovation. Commun ACM 49(7):81–85

Maglio PP, Kieliszewski CA, Spohrer JC (eds) (2010) Handbook of service science (Series: Service science: research and innovations in the service economy), Springer Science+Business Media

Normann R, Ramirez R (1993) From value chain to value constellation: designing interactive strategy. Harvard Bus Rev July–August:65–77

Ostrom AL, Bitner MJ, Brown S et al. (2010) Moving forward and making a difference: research priorities for the science of service. J Serv Res 13(1):4–36

Ricardo D (1817/2004) The principles of political economy and taxation. Dover Publications, Mineola, NY

Ruskin J (1860/2012) Unto this last and other essays on political economy. Amazon Digital Services. A Public Domain Book. Kindle Edition

Sen A (2000) Development as freedom. Anchor/Random House, New York

Smith A (1776/1904) An inquiry into the nature and causes of the wealth of nations. W. Strahan & T. Cadell, London

Soper, D, Demirkan H, Goul M (2007) A proactive interorganizational knowledge-sharing security model with breach propagation detection and dynamic policy revision. Inf Syst Front 9:469–479

Spohrer J (2012) Some books used in teaching aspects of service science. Blog post on: Service science research and education. 15 February. http://service-science.info/archives/1931

Spohrer J, Kwan SK (2009) Service science, management, engineering, and design (SSMED): an emerging discipline – outline and references. Int J Inf Sys Serv Sector 1(3):1–31

Spohrer J, Maglio PP (2008) The emergence of service science: toward systematic service innovations to accelerate co-creation of value. Prod Oper Manag 17:238–246

Spohrer JC, Giuiusa A (2012) Exploring the future of cities and universities: a tentative first step. In: Proceedings of Workshop on Social Design: Contribution of Engineering to Social Resilience, University of Tokyo, Tokyo, 12 May

Spohrer JC, Maglio PP (2010) Toward a science of service systems: value and symbols. In: Maglio PP, Kieliszewski CA, Spohrer JC (eds) Handbook of service science. Springer, New York

Spohrer JC, Maglio PP, Bailey J, Gruhl D (2007) Steps toward a science of service systems. IEEE Comput 40(1):71–78

Spohrer JC, Demirkan H, Krishna V (2011) Service and science. In: Demirkan H, Spohrer JC, Krishna V (eds) Sciences of service systems. Springer, New York

UK Royal Society (2009) Hidden wealth: the contribution of science to service sector innovation. Royal Society Policy Document 09/09. RS1496

Vargo SL, Lusch RF (2004) Evolving to a new dominant logic for marketing. J Mark 68:1–17

INNOVATION LAB
New payment methods (m-Payments)

MAINS Master, academic year 2008/2009
People and companies involved in the InnoLab:
Students: Paola Costantino, Jari Petroni and Francesco Piccioli Cappelli
Companies: Intesa Sanpaolo, Banca CR Firenze, Ericsson Telecomunicazioni, Telecom Italia and SIA-SSB
Professors: Roberto Barontini and Giuseppe Turchetti

1. The problem
Following the e-commerce era, m-commerce (mobile commerce) has recently conquered a significant segment of the market. Mobile phones have become objects of daily use that users are practically never without. Banks, mobile operators, network operators and other mediators have developed numerous payment systems that are suited to these mobile devices, with the result that today, in many cases, payments are no longer made with cash, credit cards or bank transfers, but with the comfort of the phone itself.

The business potential resides mainly in the high penetration that mobile phones have achieved. In developed or emerging countries, almost the entire population has a mobile phone. Moreover, they are usually kept close at hand, even when not used for calls or other services, thus rendering the mobile increasingly more available as a payment tool in comparison to others.

A key factor for firms investing in the m-payment market is to achieve a critical mass of customers and hence to be able to generate profits as soon as possible to compensate for the investments made to enter the sector. For this reason, the first applications to be launched were focused on the primary benefits to customers of m-payments, namely flexibility and convenience.

In this context, the workshop was conducted with the aim of analyzing and developing an innovative payment service based on the use of the mobile device as an alternative to conventional services such as cash, credit/debit cards, bank transfers, etc. Specifically, the team, having defined a perimeter of possible m-payment applications, focused on preparing a business case with the aim of providing concrete indications for potential collaborations between the project's industrial partners.

2. Work methodology
The team researched the possible areas of application relating to the m-payment project: the analysis was mainly based on screening the material collected (information on successful and unsuccessful cases) and identifying key variables in an attempt to outline the main critical success factors of each scenario.

The work progressed with a subsequent focus on a single application, considered more advantageous both economically and strategically, in order to prepare the business plan to present to a hypothetical venture capitalist.

The intention was to provide concrete input for the potential development of a business project that would open new, untapped and highly profitable market opportunities.

Following the work conducted by the team, it was decided to develop a business idea on the application of m-payments in large retail chains. For this reason, COOP Italy was involved in order to allow development of the case study.

The idea was to analyze unmet needs and services offered to COOP customers in order to implement these with the use of m-payments. Thereafter, the business potential of m-payments was assessed within the scope of large retail chains, analysing different scenarios through a set of possible services constituting the points of reference to map a business model.

3. Proposed solution

To develop the business idea, the needs of large retail chains were analyzed and were found to include reducing operating and administrative management costs, reducing congestion at checkout queues with a consequent increase in turnover, customer loyalty intended as customer satisfaction (improving the level of service) and increasing the customer base through new partners and by means of the company's image.

Similarly, customer needs were also analyzed and the results were: convenience understood as the ability to aggregate multiple functionalities in a single tool, usability with quick user-friendly services that can be enjoyed without fear of making mistakes, information transparency, or rather, access to information and visibility of transactions, a clear understanding of costs, security and protection against fraud and theft, standardisation via the scalability of services offered through the technology and standardised procedures, as well as integration with services that constitute a platform enabling the activation of other services.

The program foresaw a subsequent focus on a range of services (primary and accessory value-added) considered to be more advantageous both in economic and strategic terms.

The primary service was defined in terms of creating an internal closed circuit where payments could be made via mobiles in the contactless mode.

This service is beneficial to large retail chains since it satisfies the need to reduce cash management costs with a view to eliminating commissions and the current "war on cash", minimising management activities through virtualisation of cards and loyalty through incisive marketing initiatives.

Since a key factor for firms investing in the m-payment market is the attainment of a critical mass of customers, a closed circuit environment renders the activation of incentive mechanisms feasible – such as point accelerators, topping up phone credit, shopping vouchers and the virtual piggy bank – to reward the use of mobile payments.

Product servitisation

2

Daniele Dalli and Riccardo Lanzara

This chapter provides a discussion and an update of business practices relating to the servitisation phenomenon, or rather, the gradual but significant increase of the service dimension in customer–supplier relationships, with particular reference to manufacturing firms. The provision of services, traditionally referred to as activities that are "accessory" to sales, has long been an element pertaining to marketing and general management literature and this chapter reviews the key contributions that form part of this tradition. Following this review, recent contributions in the service science field will be considered that suggest a gradual balancing of the product (tangible) and service (intangible) dimensions in terms of the market offer. Under certain conditions, a complete reversal of the traditional perspective can be observed: from a product to a service logic. In developing servitisation, firms are no longer, or exclusively, holders of a product or manufacturing competitive advantage, but tend to increase the number and variety of sources of value offered to customers: actual services, image, design, brand, distribution and information.

2.1
What does servitisation mean?

The term servitisation was coined by Vandermerwe and Rada (1988). During the 1980s, the process of integration of services in the supply market offering was so extensive and pervasive that these elements were no longer considered as accessories

D. Dalli (✉)
Dipartimento di Economia e Management, Università di Pisa, Pisa, Italy
e-mail: dalli@ec.unipi.it

R. Lanzara
Dipartimento di Economia e Management, Università di Pisa, Pisa, Italy
e-mail: rlanzara@ec.unipi.it

L. Cinquini, A. Di Minin, R. Varaldo (eds.), *New Business Models and Value Creation: A Service Science Perspective*. Sxi 8, DOI 10.1007/978-88-470-2838-8_2, © Springer-Verlag Italia 2013

of the value added originating from manufacturing and products, but as integral and fundamental elements of this value. This trend is fully expressed in the contributions of Vargo and Lusch (2004) and Chesbrough and Spohrer (2006), who overturn the traditional perspective (the core value perceived by the customer is the product) into a true "new evaluation logic": the customer seeks solutions and hence services, and not physical objects. The customer seeks resources to translate into performance and is therefore interested in the intangible dimension of service rather than in tangible characteristics: this is the "new service dominant logic", which, according to Vargo and Lusch, constitutes the final outcome of the long process of servitisation that firms and scholars have worked on over the past 20 years.

Repositioning towards the service component would allow three key results to be obtained: (1) moving competition away from price and raising competitive barriers, (2) increasing customer switching costs, generating greater loyalty and (3) raising the level of differentiation. In addition to explanations from a strategic and competitive perspective, a number of other arguments derive from the economic structure of many sectors, such as those concerning durable goods where differentiation and competitiveness take place through the management of existing products lines (and relevant services), even prior to the sale of new products (Wise and Baumgartner 1999). Another explanation is related to environmental protection and ensuring the sustainability of processes and products: investing in the service dimension would appear to ensure positive results in this respect (Goedkoop et al. 1999).

The servitisation of manufacturing also emerges to an unprecedented extent from aggregate data: out of the 138 countries surveyed in the 2010–2011 World Economic Forum (2010) report, 98 have a level of gross national product deriving from the service economy for over 50%. In 25 countries, the ratio exceeds 70%. In Italy, it is 71%, while in Germany it reaches 69%, in the UK 76%, in the USA 77% and in France 78%.

From around the late 1980s, a debate around servitisation developed that essentially follows two main strands: manufacturing and operations literature and service marketing literature. The former develops empirical evidence and interpretive models to describe and explain how production processes are changing, and should further change, to accommodate the challenges of servitisation (Baines et al. 2009; Schmenner 2009). From this perspective, the transition from the production of products to the provision of services is essentially a challenge at the management level (Oliva and Kallenberg 2003, p. 161). This transition challenges corporate culture and management practices.

The need to intervene on organizational culture brings with it the need for action on personnel competencies and training that in the servitisation logic must be "converted" to the service logic or product and service integration. For example, moving the focus from the spot sale of standardized products to system selling (products integrated with pre- and post-sales services) enables companies to reposition from a transactional logic, oriented toward the accomplishment of a single exchange, to a relational logic, which requires an enduring interaction with and commitment to the customer. According to this perspective, the customer adopts the characteristics of both buyer and partner collaborating in a medium- to long-term perspective to develop the business and enrich it with subjective contributions. The intangible dimension of the service economy, and its eminently relational nature, require a specific mind-set (Fitzsimmons and Fitzsimmons 2008), which must be formed or reformed in the event of organisational scarcity. Initiatives such as the Mains Master and this book are important reference points on this pathway.

The impact of servitisation does not end within the boundaries of the firm at the core of the process, but has broader implications. An efficient servitisation process requires the coordination of production and service systems, the supply of accessories, components and maintenance, logistics services, etc. (Slack et al. 2004). While the product is more often created and supplied by one or a few firms, services that gradually become integrated through servitisation must be provided by several organisations (Cohen et al. 2006). This therefore requires managing the complexity arising from the integration of previously disjointed value chains where the equilibrium and the relations must be constantly adapted to suit the servitisation process. The effectiveness of the process thus progresses through the integration and coordination of people, processes and functions across distinct but interconnected value chains, providing products, components, spare parts, upgrades, maintenance services, support, consultancy and training (Johnson and Mena 2008).

The service marketing approach contributes to enhancing the service dimension in the relationship with intermediate and final users. In relation to final users, services that allow transferring value to the consumer through their integration with elements of the product have been investigated in depth. In relation to the intermediate user, analyses have been undertaken in respect of the business-to-business (b-to-b) and trade marketing areas. In these contexts, relationship and networking dynamics develop that are linked, directly or indirectly, to the North European marketing tradition (Håkansson 1982; Håkansson and Snehota 1995; Grönroos 2000; Gummesson 2002).

In the relational and network logic, value derives from the interaction between suppliers and customers. It is the relationship between customers and suppliers that generates value in the exchange and this depends largely on the ability of each of the parties to develop the business model from their own perspective: the supplier providing opportunities and resources to the customer and the customer orienting the supplier's offering towards their most relevant needs, thus "enhancing" the offer. In this process, the hard dimension of the offer (the product) progressively loses importance, since it is rigid and predefined with respect to the exchange. Meanwhile, the soft dimensions gain importance: services that are not storable, since they are intangible, and above all, highly specific to the context of the exchange and the needs of the parties involved, provide the customer and the supplier with the necessary tools to increase the value of the whole exchange.

In general, servitisation presupposes major changes in production organisation and the value chain, changes in the relationship with the customer (communication and sales) and a reconsideration of the process market value creation. Perhaps the most important evolution observed in the marketing debate concerns the role of customers in the market process and the logic with which the exchange value is determined.

Indeed, if looking at the latest developments in this field of the marketing literature (Vargo and Lusch 2004; Lusch and Vargo 2006a, 2006b), the tendency to a progressive increase of the service dimension ascribes to the customer a different and more constructive role within the market process than in the past. In the b-to-c (business-to-consumer), b-to-b and b-to-t (business-to-trade) spheres, customers are no longer regarded as those who pay money for the supply of predefined goods and services. The process that leads to the actual realisation of the market offering gives an important and constructive role precisely to the customer. This applies to both the product and service spheres. It applies to end consumers, industrial customers and commercial intermediaries. In all these contexts, the customer plays an active role that often begins very "early" if we follow the temporal dynamics with which the value chain is normally established, namely the product innovation phases. Through multiple procedures and practices, the customer is involved in product development (Prahalad and Ramaswamy 2004; Prandelli and Verona 2006), the definition of the marketing mix levers (Muñiz and Schau 2007) and also the introduction of the product to the market (Manolis et al. 2001). Post-purchase also plays a key role. Customer satisfaction arises in the consumption phase when products and services have left the supplier's domain and are physically managed by customers who interact with each other and with other institutions (Kozinets et al. 2010).

In sum, the significant literature on the servitisation theme concerns both the manufacturing dimension and thus the organisation of production within the company and between companies, as well as the marketing, sales and communication dimensions. From this perspective, firms that follow the servitisation road are forced to question their own organisational models and their relationships with other firms, and to reconfigure their relationships with customers, assigning them a constructive and participatory role. The following sections outline the key principles of servitisation in terms of its effects on marketing policies.

2.2
The service component in product marketing

"A product can be reasonably defined as the sum of physical, psychological and social satisfaction that the buyer obtains from purchase, ownership and consumption" (Peter et al. 2009, p. 142). Products are therefore very complex objects whose marketing dimensions are not only limited to the material aspects and to their physical–chemical characteristics, but extend to the world of the meanings associated with design, colour and brand, incorporating into the system the related services that include all the activities that precede, accompany and follow the sale itself. In fact, marketing, as a discipline and philosophy, has always regarded service as a key element of the offer, although moving from the concept of added service to the concept of integrated service that is part of the product itself as an inseparable and unified whole.

Some authors refer to this as the augmented product concept, which includes services such as warranty, installation, delivery and payment terms, customer service, after-sales assistance, etc. (Dibb et al. 1998).

In reality, the system of possible services that surround the product leverages mainly on the infinite possibilities that the relationship established between a firm and its customers offers, and constitutes an inexhaustible source of opportunity for the differentiation of the offering. Marketing literature teaches that there is no "undifferentiated value proposition with respect to specific standards" and hence that "commodities do not exist" (Valdani and Ancarani 2009, p. 101); all firms therefore have the opportunity to differentiate their offering. It is logical to assume that the possibility of differentiating a product depends in part on its physical characteristics: averagely complex products, such as an appliance or a car, offer significantly greater differentiation possibilities than raw materials, such as coal, or even food products, such as corn or fruit in general.

The opportunities for differentiation, in addition to the tangible component, depend on semantic and symbolic values such as brand and design, which are increasingly linked, as previously stated, to the infinite possibilities that emerge from the system of relations with the customer, which go far beyond the traditional concept of the augmented product and thus exceed the threshold of more traditional services.

On the other hand, when consumers set about purchasing a product, they search the market for a solution to their needs, their problems and their desires, and thus, on the supply side, firms position themselves on the market as problem-solving agents. It is no coincidence, as Day (2004) reported, that 63% of the top 100 firms in the Fortune ranking claim they are selling solutions rather than goods and services. Thus, products, as tools to solve problems, lose their material dimension to the point that it can be said that no one today buys or sells beds but buys or sells rest, no one buys or sells bathroom furnishings but well-being, or even buys and sells kitchens but, depending on the specific needs of consumers, fast food, slow food or even haute cuisine.

In this context, sales activities tend to break away from traditional models where the simple exchange of products/goods vs. money is central, extending to include all the activities that are necessary to satisfy the specific needs of the customer, each with his or her particular method for use and consumption. Customers tend to transform from passive subjects of the process into subjects who actively participate in identifying the most suitable solution to their problems and needs. Some authors openly refer to this in terms of consultative selling (Hanan et al. 1970; Hanan 2004). This term implies a sales process where the firm is transformed into a consultant, no longer positioning on the market as a provider but as a specialist in identifying the particular needs of consumers and hence delivering differentiated and innovative proposals.

Case study

The Satoma shop in Cecina in the Province of Pisa is a classic case of successful consultative selling. Originally born as an outlet specializing in agricultural and gardening equipment, Satoma, leveraging on clothing and equestrian equipment (breeding world-class racehorses and trotters is a local tradition), has today transformed into a point of sale for major country and casual brands such as Barbour, Timberland, Clarks, Beretta, Polo Ralph Lauren, Marlboro Classics, etc. and other articles that characterise Tuscan Maremma apparel, with a national and international clientele acquired through a series of relationships that the owner, a passionate hunter, has been able to cultivate with local and historically noble families with a long tradition in local racehorse breeding, and who are now world leaders in wine production. Today, Satoma has become a showcase for major brands to experiment their new collections and a vantage point of new tastes and emerging trends. Referring to a point of sale is, however, limiting: the owner's and the staff's main goal "is not in fact to sell, but to advise customers, even against their wishes, orientating them towards the most suitable brands or towards the brands that most correspond to their lifestyle […]. In this way, a relationship is established where dialogue and also a kind of familiarity prevails […] the exchange may end in nothing, but in most cases the customer returns" (Santini 2004).

Hence, the consultative selling service in many cases becomes an integral part of all the actions that the firm carries out with the aim of improving customer satisfaction, through an increasingly intensified characterisation of the offer.

There are also cases where manufacturing firms, as part of their strategic diversification, expand the business by extending it to services directly related to it. The value proposition in this case develops in a logic of service enrichment/aggregation, perhaps very different from the original conception of the business, but always congruent with it.

Case study

For many years, Pastificio Rana S.p.A. has operated in the production of fresh pasta, initially a typical artisan-type firm. Then it moved to industrial production, with a turnover that exceeded €300 million in 2007. It recently signed an agreement with the French chain Casinò Cafeteria to open at least 100 restaurants branded "La trattoria di Giovanni Rana" and entered the restaurant business with their own brand and product lines (www.rana.it).

Service in this case becomes a business expansion tool but always equivalent to and consistent with what Levitt (1986) defines as the concept of the general physical product that "represents and defines the minimum conditions for purchase by the customer" (Valdani and Ancarani 2009, p. 103). However, in the case of Pastificio Rana, augmented product concepts, as suggested by Levitt, are also superseded, namely, the concepts of the expected product that integrate price, warranty and delivery terms, and the augmented product that includes services that customers expect and desire beyond the standard, such as a recrafting service for footwear and leather goods or a laundry service for ties. Examples of this kind are Allen-Edmond, Hermes for the Kelly bag and Neapolitan Marinella. In fact, Rana operates in the sphere of the augmented product and thus creatively enters into the area of all opportunities that the general physical product concedes, in this case restaurant chains.

In other cases, service as a marketing tool, and from the perspective of the augmented product, becomes not only a differentiation tool but also a powerful competitive weapon for both defending from and attacking competitors.

Case study

Often the market will evolve towards consumption forms where the demand for standard products/services prevails: this is the case in the fast-food industry, with large chains such as Pizza Hut or the global phenomenon that is McDonald's. The opening of a business can obviously cause severe and adverse changes in the local supply system. In 1999, in Altamura, well known for its traditional bread, a 550 m2 McDonald's restaurant was inaugurated. Nevertheless, a small local baker, Luca Digesù, decided to open an eatery offering fresh focaccia next to the megarestaurant, attacking the multinational in its own typical fast-food sector. In this case, the differentiation strategy was clear: a product straight out of the oven against an industrial product. But the potentialities offered by the general physical product, focaccia and bread, become the augmented product, because Luca Digesù decided to enrich the basic offer integrating it with local products such as mushrooms or mozzarella, thereafter extending it to the augmented product, namely, all those activities that the physical product permits: the baker continues to make bread, thus retaining his artisan manufacturing characteristics, but alongside his original activity that was indistinct and not unique, he developed a new business, i.e., genuine quality fast food. In this way, not only did he manage to escape the dangers of price competition typical of commodities such as bread, but he also built a unique and differentiated offer enriched with customer service activity. The bakery was thus transformed into a fast-food restaurant and socialising point. Within a year and a half, despite intense advertising and promotional campaigns, McDonald's in Altamura closed down.

The www.anticacasadigesu.it case has been circulated around the world and has been commented on by the national and international press (Panorama, 28 November 2005, The New York Times, 12 January 2006, La Repubblica, 22 May 2007, Il Giornale, 4 October 2007, Corriere della Sera, 17 November 2007).

The cases presented above demonstrate that service activity is mainly expressed in the capacity to create a system of relations with customers around the base product, even if the objectives differ. These range from consultative selling in terms of clothing in the case of Satoma, to the expansion of the business, as in the case of Rana, arriving at, as in the case of the Digesù baker, quality service as a tool not only to expand the business but also as a competitive lever to attack a dominant competitor.

In all cases, however, the basic physical product is transferred and consumed, even if the competitive potential of the various subjects examined is enriched by further opportunities in the potential service system and linked to the original offer.

2.3
The primacy of the service over the product

The intense and ongoing technological innovation processes, paradoxically, have diminished the role that the physical structure of the product and hence technology can play in differentiation strategies, although they are still very important because they in fact determine the performance and functionality of the product itself. Indeed, it has to be noted how technological and innovative content, and therefore the level of performance and functionality, are often substantially similar even in products with different brands. The complexity of technological innovation is such that only a few large firms own and are able to develop certain technologies. The large investments in research and development that they require impose, in order to justify development and production costs, that they be available to all potential buyers on the market, who then incorporate them into products or processes. The scale effect, therefore, becomes dominant in given sectors, so that firms that produce technology or complex technological components and systems are typically large or are otherwise characterized by very emphatic growth processes. Consider, for example, the automotive sector where first-tier suppliers worldwide number around one hundred (Volpato 2009) and for some components, such as brake units, there are only a few manufacturers such as the Italian Brembo. This means that technological innovation allows monopolistic rent positions of a temporary type, but in the medium–long run, as a result of the processes of rapid market penetration, often trivializes and homogenizes products.[1] Consider, for example, the air-bag system. Mercedes Benz first introduced it in luxury models as an accessory, an element that strongly characterized its models, but only for a very limited period. Today it has become a standard accessory, even mandatory from a regulatory standpoint (Lanzara 2010, p. 63).

---

[1] Some authors state that "large firms do not differ significantly in terms of the basic knowledge and technology that they develop" (Zirpoli 2010, p. 41). Pavitt (1998), for example, states that firm competitive advantage is largely based on distinctive organisational characteristics rather than on distinctive technological competencies.

Firms therefore have had to put strategic differentiators in place that further accentuate the importance of not only semantic product value, but also the role that the service system can play in increasing the offer's level of uniqueness.

The binominal intangible/service component clearly changes depending on the type of good intended for the final consumer goods market, durable or non, or the industrial goods market, but in fact, empirical observation clearly highlights how today fine-tuning between these two elements has become an increasingly important weapon of competitive differentiation.

Case study

The automotive sector, which represents a pioneering field in many ways, as a forerunner of technological, organisational and management changes, has in recent years witnessed momentous transformations. Consider that for Fiat Auto in 2007, with a production value made equivalent to 100, the captive and non-captive market reached a value of 87, and the value added was therefore reduced to 23. Fiat Auto's percentage of outsourced car production and design in 2006 reached values of, respectively, 76% and 77% (Source: Fiat Auto in Volpato 2009). This demonstrates, as previously mentioned, that much of the technological product strategy is managed by the upstream supply chain and above all by first-tier suppliers. Just as there was a strong process of supplier concentration, a growth in size and a strong reduction in the number of final producers was witnessed, so there are currently 12 worldwide. The process of concentration of end players was accompanied by an extensive transformation of their competitive strategies: the carmakers of the past are today brand owners for whom the control of the network and customer care (Volpato 2009), namely customer service, is strategically important.

In b-to-b markets, service primacy is even more important (Gebauer et al 2010; Raddats and Easingwood 2010). Indeed, industrial customers tend to extensively consider the relationship with the supplier far beyond the purchase of a good or a solution in general, but also refer to the nature and quality of the system of interactions with suppliers from the problem identification phase to the definition of the solution phase, and the implementation and post-implementation service (Tuli et al. 2007; Fiocca et al. 2009).

Individual customers in industrial markets often coincide with individual segments (segments of one) and it is thus logical to assume that the solution should start from an in-depth analysis of their problems and specific needs, taking into account that the counterpart is not an individual, but a system of individuals, bound by very complex formal and informal relationships. A firm is unlikely to be comparable to another and thus the solution must be contextualised and is the result of the customer–supplier relational process. It is no coincidence that in the automotive sector the terms codesign and comanufacturing are used, highlighting the fact that the design of a component or a process technology is part of intense collaboration activities between customers and suppliers (Zirpoli 2010).

The supplier's offering is hence extended to the concept of augmented products that include services that the customer expects and desires beyond the standard. A machine tool buyer therefore also acquires a system of services closely linked to it, such as consultancy and technical cooperation in relation to the identification of appropriate technology, its installation, organizational and financial assistance. Anyone buying a computer is not just buying hardware, but also acquires the redesign or modification of information systems, process analyses and the development of application software (Fiocca et al. 2009, p. IL8). Those buying a fuel injector often acquire a complex device specifically designed for that type of engine and are thus not only buying a design that is frequently implemented in collaboration with the supplier but also acquire the supplier's testing and debugging capabilities.[2]

The computer industry, with particular reference to personal computers, has some interesting cases where the service activity and therefore the development of applications and client-specific solutions become the b-to-b operator's core business, while hardware design and production are delegated to third parties. This is understandable when considering that the PC is a product derived from the integration of modular components, which is greatly facilitated by standard interfaces between the subsystems that compose the product (Sturgeon 2002). In this way, the product has to all effects become part of the commodity sector where competitive strategies prevail aimed at maximum efficiency and where cost reduction becomes essential. IBM, for example, has sold the entire production of personal computers to China's Lenovo, focusing exclusively on offering IT solutions and consultancy, thus substantially changing its value proposition. Accordingly, IBM today is no longer a manufacturing business, but should instead be considered as an advanced service company: turnover in 2008 was $103 billion, of which $80 billion (78%) came from selling services and software.

Case study

IBM's website states: "The mission of IBM Global Business Services is to collaborate effectively with clients and together face their most complex business issues. We apply our business insight to developing innovative solutions that produce tangible and measurable results, whether it be to design and implement a new service following the redesign of a business model for the sales cycle, or to revolutionize a car insurance business model with the introduction of innovative technologies, or to become leader in the provision of logistics for the supply chain. We work with our clients to identify the degree of change that best fits their needs and translates into sustainable results. We put together the best of IBM and our business partners to implement the change and optimize our customers' performance."

---

[2] Debugging refers to the process of the identification and correction of design and process errors.

Furthermore, "The company's business model is built to support two principal goals:

1) helping clients succeed in delivering business value by becoming more innovative, efficient and competitive through the use of business insight and information technology (IT) solutions;
2) providing long-term value to shareholders." (www.ibm.com).

The cases of the automotive and computer industries, with particular reference to Fiat Auto and IBM, are thus two significant examples of how service prevails over the product: in both cases, to a different but still significant extent, product design, development and production is delegated to third parties. The core company in the Fiat case centres the competitive strategy on the brand and on commercial network management and valorisation, focusing on maximising customer care. In the IBM case, the product disappears completely from the firm's value chain, where instead the core business is constituted by customer services. In both cases, however, as in the previous ones, there is still a physical transfer of the product since the end user buys the product and is able to use the services linked to it, although, as in the IBM case, products and services are acquired by two different operators.

Conversely, the emphasis shifts from the sale of products to the sale of use in the case of Rolls-Royce: "here, rather than transferring ownership of the gas turbine engine to the airline, Rolls-Royce delivers power-by-the-hour" (Baines et al. 2007, p. 1543). Another interesting case comes from the telecoms industry. In this sector,[3] although characterized by very intense and fast technological dynamics, the hardware, meaning the physical network, both via telephone and fibre, tends to be standardized. Therefore, even in this case, as in the personal computer industry, given the technological content, the physical product becomes a commodity characterised by strong price competition: software applications thus become a powerful lever of differentiation. The very high levels of competition in this arena lead some firms to further develop their differentiation strategy aiming not only at consultative selling activities, but to maintain ownership of the customer network and its applications, becoming their managers in toto: the customer purchases the use of the network, i.e., so-called managed services.

Case study

Sweden's Ericsson Marconi, worldwide leader in network technologies, uses an entirely informal marketing approach compared to the technological/centric tradition that is typical of companies developing highly innova-

---

[3] These considerations arise from the collaboration between the author and Stefano Coiro, Head of Business Development and Sales Support at Ericsson Italy (Lanzara and Coiro 2005).

tive products and in all respects it adheres to the consultative selling philosophy. The starting point of the process of developing a new technological proposal is in fact an analysis of the needs of the end customer and is
achieved by following an inverse logic (backward process) from the market
to the choice of technology through the phases shown in Fig. 2.1 (Lanzara
and Coiro 2005).

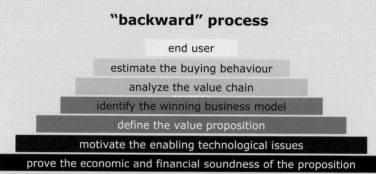

"backward" process

end user

estimate the buying behaviour

analyze the value chain

identify the winning business model

define the value proposition

motivate the enabling technological issues

prove the economic and financial soundness of the proposition

Fig. 2.1 The validation process of the technological options at Ericsson Marconi
(Lanzara and Coiro 2005)

In sum, proceeding with the following steps:

1) Study of end-user purchasing models and motivations (What do customers want?).
2) Identification of the value chain aimed at satisfying the purchasing
   model (How to satisfy end customers).
3) Identification of the value chain's best organization and operation model
   (winning business model).
4) Selection and definition of the proposed technology (value proposition).
5) Estimation of the potential value for the industrial customer (e.g., Fastweb) of the proposed new technology.
6) Estimation of the cost of the economic–financial investment by the industrial customer.

Marketing at Ericsson Marconi, therefore, not only has the task of analysing the
market and studying the purchasing behaviour of end customers, but also plays the
important role of selecting the trajectories and the various technological solutions,
following an appropriate marketing concept approach that places customers and their
satisfaction at the centre of the company processes through a value proposition that
is congruent with their needs.

It is no longer the garment that is most congruent with the lifestyles of its customers, as proposed by Satoma, but a complex technological product that constitutes
the tool with which Ericsson's direct customers can build their best value proposition
and that best addresses the needs of the end customer.

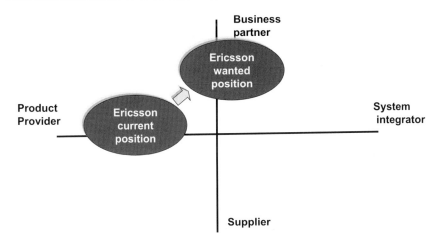

Fig. 2.2 The Marconi Consultative Approach (internal Ericsson Marconi 2005 documentation)

However, Ericsson Marconi does not limit itself to consultative selling. It also proposes itself as the customer network manager, becoming a managed service provider who plans, optimizes and develops the network, its services and applications, ensuring technological development and ensuring its maintenance.

Ericsson Marconi thus defined a new business model, with customer service at the core of its competitive strategy, configured in all respects as a partner (Fig. 2.2) in a process that leads to the integration of the respective value chains.

The Ericsson Marconi case clearly shows how the object of exchange is ultimately the use of technology, namely, what some authors called the product-service system (PSS) (Baines et al. 2007). The identity of the product evolves towards forms where the physical component is completely inseparable from the system of services associated with it. Similarly, service includes the product, thus forming an inseparable system (Morelli 2003). In this respect some authors claim: "The key difference between the 'old service model' appears to be that while the former concerns providing services that support the product, the new service model provides services that actually support the customer" (Mathieu 2001).

While the traditional concept of service referred to its integration with the product, today the reverse concept prevails: it is service that integrates the product, the system of customer/supplier relations becomes the main object of the value proposition and it is precisely the system of relations that this defines and develops. Hence, the customer, by interacting with the supplier, interprets and helps cocreate value (Michel et al. 2008, p. 58). Some authors note, "Value is realized, not released, because value is not for exchange but rather in use. The value-creating process is truly the co-creation of value among providers and customers [...] value is not defined by a firm alone" (Michel et al. 2008, pp. 50–52). It is therefore logical to consider that a new form of innovation has been established that translates into new and different forms of use of the potentialities offered by technology while the offer evolves towards a complex set of products, services and customer relationships, "Value is not produced and then transferred to the customer, but rather value is co-created by a customer, who has recognized some potential value in actualizing the service that an offering provides to him" (Michel et al. 2008, p. 65).

2.4
The reaction of the product to the challenge of the service-dominant logic: servitisation

This chapter began with the observation that a real revolution is underway in management theory and practice. A revolution that leads us to consider service at the core of the value perceived by customers. According to the service-dominant logic (Vargo and Lusch 2004; Lusch and Vargo 2006b), firms must take note that the customer seeks service and not products, and should therefore provide services, or rather, ways of appreciating the value of their resources linked to the intangible dimension of the purchasing and consumption process. This revolution leads the service logic to take on a "dominant" role with respect to a traditional approach that is substantially anchored to the product and to production. This approach allows identification of the drivers of value in the physical characteristics of the product, which are in turn anchored to design and manufacturing.

As we have tried to explain on a theoretical and empirical level, the service-dominant logic can be seen as the culmination of a long and articulated process that in the last twenty years has affected both manufacturing processes and finished products and goes by the name of servitisation (Vandermerwe and Rada 1988), namely, the joint offering of products and services, with the latter gradually assuming increasing importance. Through this process the role of the product has not been depleted, but has changed form and content and has evolved into a new dimension, which is precisely the product-service dimension as described in the preceding pages.

This is the process through which a real service science has formed (Maglio and Spohrer 2008; Spohrer and Maglio 2008) where the service dimension is integrated with that of the product, while operations and information systems are integrated with the marketing dimension. This theoretical and managerial path, as already emphasized by Spohrer and Kwan in their contribution to this text, requires interdisciplinary approaches and a willingness to review the manufacturing and management processes, which cannot be simply abandoned. This willingness is reflected in the evolution of several large groups, some of which are discussed in the preceding pages, that are leading this trend on both national and international levels.

References

Baines T, Lightfoot H, Benedettini O, Kay J (2009) The servitization of manufacturing: a review of literature and reflection on future challenges. J Manuf Technol Manag 20(5):547–567

Baines TS, Lightfoot HW, Evans S et al (2007) State of the art in product-service systems. Inst Mech Eng Proc 221(Part B):1542–1552

Chesbrough H, Spohrer J (2006) A research manifesto for services science. Commun ACM 49(7):35–40

Cohen MA, Agrawal N, Agrawal V (2006) Winning in the aftermarket. Harvard Bus Rev 84(5):129–138

Day GS (2004) Aligning the organization to the market. J Mark 58(October):43–57

Dibb S, Simkin L, Pride WM, Ferrell OC (1998) Marketing concepts and strategies. Houghton Mifflin, Orlando

Fiocca R, Snehota I, Tunisini A (2009) Marketing business to business [in Italian]. McGraw-Hill, Milan

Fitzsimmons JA, Fitzsimmons MJ (2008) Services management: operations, strategy, information technology. McGraw-Hill, London

Gebauer H, Paiola M, Edvardsson B (2010) Service business development in small and medium capital goods manufacturing companies. Managing Serv Qual 20:123–139

Goedkoop M, van Halen C, Te Riele H, Rommens P (1999) Product service systems: ecological and economic basics. Ministry of Housing, Spatial Planning and the Environment, Communications Directorate, The Hague

Grönroos C (2000) Service management and marketing. A customer relationship management approach. Wiley, Chichester

Gummesson E (2002) Total relationship marketing: marketing management, relationship strategy and CRM approaches to the network economy. Butterworth Heinemann, Oxford

Håkansson H (1982) International marketing and purchasing of industrial goods. Wiley, New York

Håkansson H, Snehota I (1995) Developing relationships in business networks. Routledge, London

Hanan M (2004) Consultative selling. Amacom Books, New York

Hanan M, Cribbin J, Heiser H (1970) Consultative selling. American Marketing Association, New York

Johnson M, Mena C (2008) Supply chain management for servitised products: a multi-industry case study. Int J Prod Econ 114(October):27–39

Kozinets RV, de Valck K, Wojnicki AC, Wilner SJS (2010) Networked narratives: understanding word-of-mouth marketing in online communities. J Mark 74(2):71–89

Lanzara R (2010) L'Evoluzione della filosofia progettuale dei prodotti industriali. In: La scuola di Riccardo Varaldo. Relazioni personali e percorsi di ricerca. Pacini, Pisa

Lanzara R, Coiro S (2005) Il Marketing delle nuove tecnologie. La sfida dei prodotti e dei mercati inesistenti. Atti del Convegno Nazionale SIMktg, Trieste

Levitt T (1986) The marketing imagination. Free Press, New York

Lusch RF, Vargo SL (2006a) Service-dominant logic: reactions, reflections and refinements. Mark Theory 6(3):281–288

Lusch RF, Vargo SL (2006b) The service-dominant logic of marketing: dialog, debate, and directions. M.E. Sharpe, Armonk

Maglio P, Spohrer J (2008) Fundamentals of service science. J Acad Mark Sci 36(1):18–20

Manolis C, Meamber LA, Winsor RD, Brooks CM (2001) Partial employees and consumers: a postmodern, meta-theoretical perspective for services marketing. Mark Theory 1(2):225–243

Mathieu V (2001) Product service: from a service supporting the product to a service supporting the client. J. Bus Ind Mark, 16: 39–58

Michel S, Brown SW, Gallan AS (2008) service-logic innovations: how to innovate customers, not products. Calif Manag Rev 50(3):73–92

Morelli N (2003) Product-service systems, a perspective shift for designer: a case study – the design of a telecentre. Des Stud 24(1):73–99

Muñiz AM, Schau HJ (2007) Vigilante marketing and consumer-created communications. J Advert 36(3):35–50

Oliva R, Kallenberg R (2003) Managing the transition from products to services. Int J Serv Ind Manag 14(2):160–172

Pavitt K (1998) Technologies, products and organizations in the innovating firm: what Adam Smith tells us and Joseph Schumpeter doesn't. Ind Corp Change 7(April):433–452

Peter JP, Donnelly JH, Pratesi CA (2009) Marketing [in Italian]. McGraw-Hill, Milan

Prahalad CK, Ramaswamy V (2004) The future of competition: co-creating unique value with customers. Harvard Business School Press, Boston

Prandelli E, Verona G (2006) Collaborative innovation [in Italian]. Carocci, Rome

Raddats C, Easingwood C (2010) Services growth options for B2B product-centric businesses. Ind Mark Manag 39(8):1334–1345

Santini F (2004) Lo sviluppo di Cecina come polo commerciale. Un mini business di successo: il caso Satoma di Marco Rindi. Tesi di Laurea, Dipartimento di Economia Aziendale, Università di Pisa, 2003–2004

Schmenner R (2009) Manufacturing, service, and their integration: some history and theory. Int J Oper Prod Manag 29(5):431–443

Slack N, Lewis M, Bates H (2004) The two worlds of operations management research and practice: can they meet, should they meet? Int J Oper Prod Manag 24(4):372–387

Spohrer J, Maglio PP (2008) The emergence of service science: toward systematic service innovations to accelerate co-creation of value. Prod Oper Manag 17(3):238–246

Sturgeon TJ (2002) Modular production networks: a new American model of industrial organization. Ind Corp Change 11(3):451–496

Tuli KR, Kohli AK, Bharadway SG (2007) Rethinking customer solution: from product bundles to relational process. J Mark 71(May):1–17

Valdani E, Ancarani F (2009) Marketing Strategico: Manovre e Strategie di Marketing [in Italian], vol 2. Egea, Milan

Vandermerwe S, Rada J (1988) Servitization of business. Eur Manag J 6(4):314–324

Vargo SL, Lusch RF (2004) Evolving to a new dominant logic for marketing. J Mark 68(1):1–17

Volpato G (2009) La Supply Chain nella Strategia Aziendale: dalla competition alla co-opetition. Atti del Convegno ADACI, Bologna, 19 November

Wise R, Baumgartner P (1999) Go downstream. Harvard Bus Rev 77(5):133–141

World Economic Forum (2010) The Global Competitiveness Report 2010–2011. http://www.weforum.org/en/initiatives/gcp/Global%20Competitiveness%20Report/index.htm

Zirpoli F (2010) Organizzare l'Innovazione. Strategie di Esternalizzazione e processi di apprendimento in Fiat Auto. Il Mulino, Bologna

INNOVATION LAB
Product servitization

MAINS Master, academic year 2009/2010
People and companies involved in the InnoLab:
Students: Marco Fontana, Matteo Gloyer, Laura Prete and Francesco Tantini
Companies: CDC, Coop Italia, IBM Italia and Telecom Italia
Professors: Roberto Barontini and Giuseppe Turchetti

1. The problem
Servitisation is a concept that was first introduced by S. Vandermerve and J. Rada in 1988. The authors define it as "market packages or bundles of customer focussed combinations of goods, services, support, self-service and knowledge", or rather, the "servitisation" process, understood as the transition from the logic of product mix to the logic of the above-mentioned components where services form the dominant part.

Following this logic, products are no longer sold as such, but as a service distribution vehicle (the servitisation process in fact entails various stages: beginning with a product that offers services arriving at a service that has an adjoining product).

Having effectively defined the process of servitisation, a historically manufactured product was identified in the lab that could be subjected to the servitisation process: the "television set".

Technological innovations introduced by TV manufacturers have driven the extensive evolution of the TV set. The Internet-connected TV is a recent development in TVs that are sold on the market today: the producer provides an ad hoc platform in which the service provider can ask to include certain services. The innovative logic introduced in the lab was based on the possibility that a retailer could use this feature to sell personalised services through the manufacturer's platform.

Specifically, the case analysed the possibility of Coop Italy selling on-line shopping services for products not found in-store through new features introduced in TV sets connected to producers.

The idea was to create a platform of services implemented on an Internet-connected TV with the aim of rendering the company–customer relationship more stable and enduring over time and space, namely to "bring the store to the customer's home". The scenario would create a new business model that, through the exploitation of TV sets sold by Coop Italy, could strengthen the relationship with its partners as well as gain significant competitive advantage and thus obtain new sources of revenue through the introduction of new and increasingly relevant services.

Television is still the mainstay of family gatherings and, according to ISTAT data, is present in 96.1% of Italian homes. The technological development that has led to the development of the Internet-connected TV is consistent with the fact that 47.3% of households have Internet access and search for information on products and services through the Internet.

According to this analysis, the opportunity for the retailer, Coop Italy in this case, is to implement a new business model consistent with the new market logic as well as with the changing needs of consumers increasingly looking for innovative services.

2. Work methodology

In the first phase of the work a literature review was conducted on the previously presented concept of "servitisation", understood as the phenomenon of the demise of the centrality of the product and the subsequent transition to a mix of products, services, support, self-service and knowledge. Different levels of servitisation were emphasised in this review, starting from the pure manufactured product through various intermediate products with a more or less important service component attached, to a maximum level where the product is merely a service delivery tool, bringing to light the motivations that drive companies to servitisation and the diffusion of the phenomenon at global level.

In a second phase, knowledge of the concept of servitisation was broadened by searching for specific examples, starting with "classic" examples such as Rolls-Royce's "Power by the Hour" programme, which from the sale of aircraft engines has progressed to selling the power of these engines based on time service contracts that include managing and maintaining the engine. This research has shown that the phenomenon is by now consolidated in the b-to-b sector, whilst still embryonic in the b-to-c sector.

The third phase was characterized by brainstorming sessions to find a product that today is sold as such, and on which a servitisation strategy could be implemented. As previously mentioned, the product that was identified was Internet-connected TV, an innovative product that connects the world of traditional television with the Internet, providing access to various multimedia platforms, information services and social networks.

To better frame the context, research on the social context and the market for televisions and related products such as PCs, tablet PCs and smartphones was undertaken. This research provided the basis for identifying possible interaction scenarios between the manufacturer, service/content provider and distributor.

In a further brainstorming session, various types of services were investigated in order to isolate those in line with the chosen scenario. Detailed interviews with representatives of major Internet-connected TV manufacturers were conducted to form the basis for the development of a service package as well as defining the aspects of actuating a user interface to introduce into the connected TV.

In the final phase of the project, the actual business case was developed, assessing the costs and possible sources of revenue associated with the services. These include both revenue through direct sales and advertising space within the platform as well as indirect effects due to customer loyalty and expanding the range of products sold. The business plan developed in this analysis demonstrated the great potential of the project and highlighted the servitisation process that leads to a progressive increase in revenue from services while the product becomes a mere "commodity".

3. Proposed solution
Although the TV has always been considered a manufactured product, service components are increasingly evolving because of the rapidly changing technological context. Thus, since the TV market is ever-expanding due the growing percentage of TVs connected to the Internet (also due to the gradual switch-off of the analogue signal), as well as the favourable economic–social environment, the project foresaw the possibility of exploiting the connection potential from the retailer's point of view, "servitising" TVs through offering innovative services in addition to traditional services.

The idea was to create a platform of services implemented on a connected TV with the aim of making the firm–customer relationship more stable and enduring, claiming to "bring the store to the customer's home". This idea would benefit not only retailers but also customers and manufacturer: the customer would actively participate in every phase of the product life cycle and the manufacturer would further increase the number of TVs sold, in addition to the ability to extending this type of servitisation to other products. The plan foresaw, following the business case proposal in 2010, agreements with producers, the preparation of the platform in 2011 and, finally, in 2012, market entry of TVs connected to retailers.

The hypothesis of services to be conveyed through the connected TV included: pre- and post-sales services such as virtual assistance in choosing consumer electronics or in post-sales management; sales services, for example in relation to products not available at the physical point of sale or for consumer electronics; information services such as information on the world of Coop and on the origin of products.

From a temporary boom in sales of connected TVs between 2008 and 2010, also caused by the gradual switch-off of the analogue signal, the percentage is expected to increase to reach 90% in 2015. Retailer-branded connected TVs will be launched on the market in 2012, while in the following two years an estimated 90% of TVs sold by the retailer will incorporate the new service platform. In subsequent years, every connected TV sold will include the new service platform.

In the first year of market entry, a €50 discount on the purchase of the product will be offered, which should achieve a market share of around 10%. To encourage the use of the services, a bonus is foreseen for those who exceed a predetermined level of spending on the connected TV channel. Anticipating an ever-increasing percentage of people who purchase and receive a bonus, growth in both turnover and margins are expected in the years 2012–2015.

In estimating the discount and marketing expenditure as transaction costs, the cost of the discount is expected to be balanced in the first year with a contribution from the manufacturer, while increasing margins are expected from both advertising revenue and from digitally transmitted flyers. At this point, net of the platform maintenance costs, after the first year of operating at a loss, a return to profit is foreseen from the second year onwards. Even from a financial standpoint, with a multi-year estimate the project is shown to create value. The most important thing is that while the share of revenues from the sale of TVs remains almost constant, the percentage of revenues from services steadily increases over the years. The entire innovative servitisation process in a mature industry creates new sources of revenue for the retailer, constitutes a stronger bond with customers and, ultimately, by creating a new business model, allows replication of the servitisation path for other types of devices and products.

Business model innovation paths

3

Henry Chesbrough, Alberto Di Minin, and Andrea Piccaluga

This chapter explains the business model concept and explores the reasons why "innovation" and "innovation in services" are no longer exclusively a technological issue. Rather, we highlight that business models are critical components at the centre of business innovation processes. We also attempt to describe the characteristics of a business model and what a "good model" should achieve. Moreover, we discuss why firms in the service sector too often fail to innovate their business models as required by market situations. In the second part of this chapter, starting with the description of a problem or a new opportunity, we present a framework to help innovation teams develop a structured brainstorming path in order to create a new business model.

3.1
Introduction

In the first two chapters of this book, the centrality of the combination of components into a single offering of products and services was emphasised. This combination, which more and more frequently characterises the offerings of firms of all sizes and sectors, is aimed at overcoming the so-called "commodity trap". This concept is

H. Chesbrough
Garwood Center for Corporate Innovation, University of California, Berkeley, CA, USA
e-mail: chesbrou@haas.berkeley.edu

A. Di Minin (✉)
Istituto di Management, Scuola Superiore Sant'Anna, Pisa, Italy
e-mail: a.diminin@sssup.it

A. Piccaluga
Istituto di Management, Scuola Superiore Sant'Anna, Pisa, Italy
e-mail: a.piccaluga@sssup.it

L. Cinquini, A. Di Minin, R. Varaldo (eds.), *New Business Models and Value Creation: A Service Science Perspective.* Sxi 8, DOI 10.1007/978-88-470-2838-8_3, © Springer-Verlag Italia 2013

presented and discussed by Henry Chesbrough (2011) in his book *Open Service Innovation*, which explains how the new concept of product–service combination is obtained. In particular, Chesbrough argues that, from the open innovation perspective, this combination should be configured and interpreted as a real platform where contributions from customers and suppliers converge.

In this respect, as Dalli and Lanzara emphasise in Chap. 2, and as Merli will articulate in the following one, business model innovation, the subject of this chapter, is central. The reader will be introduced in these pages to a way of changing the current business model, which is precisely designed to enhance a firm's offerings, with external contributions that render it better equipped to selectively and effectively integrate the qualifying service components.

The aim of this chapter is therefore to provide the reader with an operational tool to guide reflection about changing the business model in a specific firm, both incrementally and radically, and finally reach a combination of new products and services. A path is presented, in the form of a journey in eight stages, starting from the definition of a problem up to reaching the new business model goal.

As pointed out by several other contributions in this book, it is important that firms think about innovating their business model starting from the assets at their disposal without hesitating to get involved in changes, even significant ones, in their way of doing business.

Thereafter, a structure will be laid out, a method for reflecting on how to formulate the search for change. The concept is that readers and potential adopters – whether managers or scholars – become familiar with the methodology and then try to apply it to a variety of issues that may emerge in the course of their work or in the analysis of other firms' business models. The sources of change are frequently on the outside, in the form of new problems or opportunities, pushing firms to think and question their own business-as-usual attitude, which may lead to lock-in phenomena. In these cases, tools are required that guide both the reflection and the path towards change.

The contents of this chapter were structured by the authors during several corporate training sessions, in different contexts. This tool was adopted, for example, in company retreats where business executives and managers were encouraged to reflect on the evolution of the competitive scenarios affecting their businesses. In some cases, customer business developers were also involved, with particularly interesting results. Another applicative example is the Business Model Innovation module of the Scuola Superiore Sant'Anna's master's course in innovation management (MAIN), particularly in the 2009 and 2010 editions, during which the structure of the methodology in eight phases was adopted to present several theoretical tools and at the same time apply them to the contents of the innovation labs within the course. The results of these are illustrated in the boxes in this book.

The reflection and involvement of students, managers and consultants adopting the proposed methodology described in these pages should ideally be undertaken with alternating phases of preliminary teamwork in parallel with plenary sessions where the team's progress is presented in brainstorming sessions.

However, prior to introducing the concrete steps that need to be taken on this path, we will recall some of the theoretical issues addressed in other chapters of this book since they provide valuable starting points for fundamental business model innovation processes.

In fact, over the past two decades, companies of all sizes have found themselves having to increase their research and development (R&D) expenditure to meet the emerging competitive challenges. The continuous integration and specialisation of research tools and technologies is also at the root of the problem of the so-called technology paradox: a phenomenon whereby the technologies adopted by a firm become more complicated and expensive the more numerous they are; also the management of innovative projects becomes more complicated and expensive and hence the higher the risk of technologies becoming obsolete and losing market value, particularly in light of the rapid pace of technological innovation today.

The intensification of global competition and the extensive exchange of competencies between countries and companies have also increased the rate of competition in the ideas marketplace, which now takes the form of an international network of individuals and knowledge centres that go far beyond the boundaries of a single company. Firms, however, have the opportunity to tap into this network to seek original solutions and to establish the division of labour on a global scale, even for more innovative projects. For some industries, this process of specialisation and collaboration has gone far beyond the experimental stage and is now a necessary source of a very established business practice. Consider, for example, collaborations between pharmaceutical companies and biotech start-ups. Large pharmaceutical companies often obtain research results, at different stages of advancement, developed outside of their laboratories. Research is often conducted in university laboratories or other public research centres, perhaps distant from the market logic, but then it becomes more attractive after the work of start-up firms or academic spin-offs that incrementally add value to the results, leading to successive development stages.

Other industries, such as information technology or electronics, have also been affected by these dynamics, with the result that large companies have moved away from a "closed" innovation model focused on the role of their large R&D centres and have "opened" themselves up to external contamination. The term "open innovation" (Chesbrough 2003) is used precisely to describe this phenomenon, which is affecting almost all industrial sectors (Fig. 3.1). Talking of open innovation leads directly to addressing the issue of business model innovation (Chesbrough 2008), in both the product and service sectors.

The concept of open innovation can be summarised in the following statement: "I must always think that the best people to carry forward an innovative process do not work for my company […] they are definitely somewhere else out there!"

Why is it necessary to think of new business models when it comes to open innovation? Because we need to start thinking differently of the firm, reviewing, as proposed by Merli in Chap. 4, the processes from a different perspective, and considering its product offer as a platform where internal competencies and those of suppliers and customers can be combined. This requires specific rules of engagement and partnership, pricing systems, divergent incentives and therefore new business models. Extensive creative effort is required to obtain new business models,

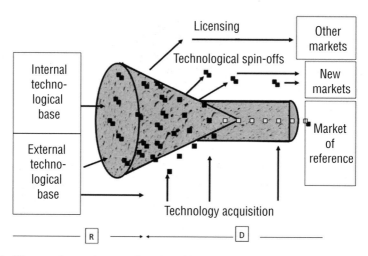

Fig. 3.1   The open innovation paradigm (modified from Chesbrough 2004)

and with the path proposed in these pages, we would like to help readers with their strategic thinking.

We could more formally state that the open innovation process is characterised primarily by a combination of an internal technological base with an external technological base. In addition, the new solutions, new products and new services resulting from this process are not necessarily brought to the market by the innovator but can also be transferred to other business entities, which will then be responsible for combining them with their complementary assets and marketing them.

From a strategic point of view, open innovation consists in the following two fundamental activities:

1) acquisition of external knowledge, both at the beginning, i.e., in the fundamental research phase, and in the most downstream stages of the innovation process;
2) valorisation of the innovative results on the technology market, entrusting to others the responsibility of commercialising new products and services.

Combining these two strategies and activities is fundamental in the service economy that this book is presenting, in particular to overcome the commoditisation trap through new business models.

Figure 3.2 represents the structure of this chapter and also presents the stages along the proposed path. We start with the definition of a problem to work on – which is also, often, an opportunity – and then discuss the development of the scenarios that could somehow affect and determine the implementation of a new model. "Enabling" technologies will then be considered that allow proposal of a solution for the identified problem. In the second part of the chapter, we explore the need to combine external technologies with internal solutions on the market, through alliance strategies that can lead to effective offerings for new product ideas. Subsequently, considering both internal and external resources, we will arrive at a new

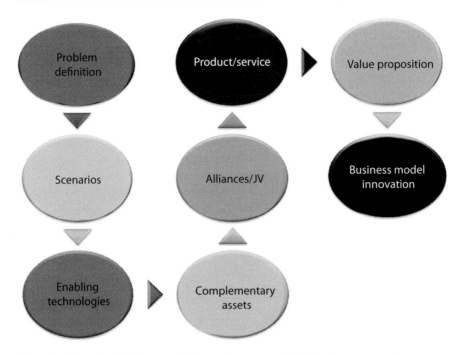

Fig. 3.2   The path to business model innovation presented in this chapter

product/service configuration that we will present in the third part of the chapter through a new value proposition.

3.2
The destination: a new business model

We now recall some fundamental concepts about what a business model is. Prior to setting off on a journey towards change, a logical framework of reference is needed that is able to represent, schematise and operationalise the key variables at play. A business model is above all a frame of reference, a paradigm, something that logically explains how to get from point A (an idea) to point B (business results) through the application of certain technologies, elaborating the correct execution of specific strategies, a new solution that combines products and services. Effective business models are like recipes: they provide accurate information, but require careful and expert execution and, as often happens, the same recipe produces very different results depending on the chef, and these are not always guaranteed. Risk is a key component in implementing a business model. Indeed, the presence of risk justifies the economic reward of the entrepreneur who decides to face an uncertain situation, creating a (new) business model.

In summary, we can define a business model as a logical scheme that links ideas, technologies and economic performance. It explains how, through entrepreneurial effort, an organisation can transform its potential into new value. In particular, a complete business model should have the following six objectives:

1) Specify a value proposition. How can I generate new value for users?
2) Identify a market segment. Who are these potential users?
3) Identify the value chain beginning from raw materials, competencies and the various manufacturing phases, up to the end customer.
4) Describe the position of the firm within the value chain and identify the complementary assets that are needed by the firm to contribute to this path.
5) Specify the mechanisms of revenue generation in order to cover the business costs and obtain profits.
6) Specify the competitive strategy that differentiates the new offering from that of competitors and attempt to forestall.

The final schematisation of the business model is fundamental to undertake an innovation path. We specifically refer here – amongst the various schemes available today – to the business model canvas proposed by Alex Ostelwarder (2010).

The six functions of a business model can be represented in the schematisation illustrated in Fig. 3.3, whose function is to clearly explain what the firm proposes to sell (WHAT), who the customer base is (TO WHOM), how to obtain the combination of competencies and inputs required (HOW) and finally the economic and financial equilibrium to be reached (HOW MUCH). In the illustration, these elements are placed in relation to each other and therefore depict a logical scheme that can be used as an outline in deliberating the business model innovation path. In the course of the eight stages described below, old and new combinations will be questioned and proposed, without relinquishing a holistic view of the overall picture in

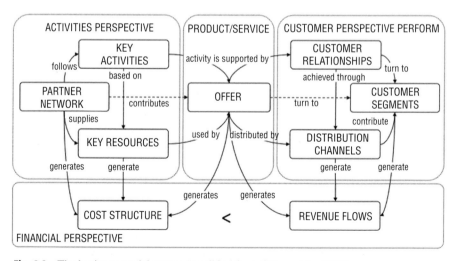

Fig. 3.3   The business model canvas (modified from Osterwalder 2009)

all its components: What, Who, How, How Much. At the end of the business model innovation path, this diagram will be retrieved and, once wholly completed, it will represent the real codified output of the path.

3.3
Problem setting

The first step in this innovation path is to define the problem to set out from. It is crucial to identify a starting point for the business model renewal path. For example, what is slowing the growth of our company? Where do we risk losing advantage with respect to the competition? What opportunities can we use for greater differentiation? A good starting point is fundamental and thus requires identifying, together with the individuals selected to conduct the exercise, a number of possible ideas and suggestions, perhaps using one of the several brainstorming techniques that are normally used. External stimuli could be useful to formalise this step, for example, a top management proposal or the most urgent firm priorities. In this situation it is useful to try and absorb and reinterpret the opportunities and specify and operationalise the selected issue. The use of brainstorming techniques at this stage is absolutely central since the various working groups involved, working in parallel on the same issue, could reinterpret the particular issue in different ways while offering their own original interpretations. During this problem definition stage it is essential that various dimensions emerge that could be useful to the analysis and to the risks that the firm may be exposed to if the problem remains unresolved, and of course the benefits that may arise in grasping the new opportunity.

3.4
Scenario planning

Once the opportunity has been clearly identified and described, it is essential to frame the evolution of the business model in a broader context, namely, where the company is repositioning to and what characterises the economic sector in which it operates. Of course, the complexity and richness of the evolution of the industry concept go far beyond the scope of this chapter, but it is sufficient to say that the study of the scenario is aimed at characterising the reference variables and helping management discuss how to develop the areas in which to compete. A useful and logical schema for scenario planning is offered by Garvin and Levesque (2006), through a path of eight easy steps as depicted in Fig. 3.4.

Key focal issue. The definition of a scenario starts from a basic problem: an issue that has profound uncertainty but is particularly relevant to the business. A scenario can have a time horizon of several months or several years, depending on the key focal

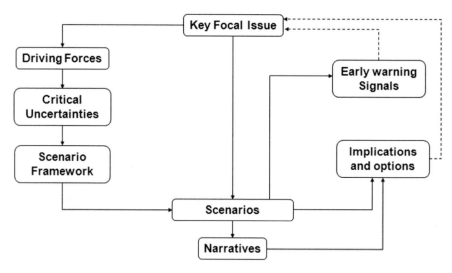

Fig. 3.4 Defining the phases of a scenario (Garvin and Lavesque 2006)

issue. Clearly, the choice of issue must be relevant to the business model innovation process.

Driving forces. What factors have an impact on the scenario? If the chosen issue is ambitious enough, it will soon become clear that the factors determining the dynamics are manifold and very complex: these too are far from predetermined. The selection of the driving forces affecting the focal issue require operating according to a brainstorming logic, identifying as many variables as possible, while leaving ample room for discussion on their relevance.

Critical uncertainties. Identifying the two most important variables. Every model must be able to define some drastic simplifications. This is a critical step to substantiate the discussion, and create a powerful presentation tool. Subsequently, once the various factors that can influence the scenario have been identified, two of the factors considered most relevant for the identified scenario must be converged on and should be as orthogonal as possible (i.e., uncorrelated). This choice must also take into account the fact that, for the chosen time horizon, these variables are actually entirely random and out of the firm's control. Either qualitative or quantitative variables can be considered.

Scenario framework. The intersection of the two variables forms a 2×2 matrix. The two extreme values that one of the two selected variables can take will be positioned on the x-axis while the extreme values of the other variable will be positioned on the y-axis.

Scenarios. The four boxes of the matrix represent four alternative worlds. They are achievable depending on the development in one way or the other of the two critical variables. Assigning a name to these four situations representing the four scenarios is useful for communication purposes.

Narratives. At this point, it is possible to focus on the description of these four possible worlds.

Implications and strategic options. Once the contents of these various situations have been described, it is necessary to return to the business model to be pursued and therefore think about the strategic implications and options available depending on the projection in one direction rather than another. Reflecting on which of these four scenarios may be more desirable can also lead working groups to reflect on the original problem, requalifying or identifying possible solutions of business alternatives thus far not taken into consideration.

Early warning signals. To conclude this exercise, it may be necessary to return to the analysis of the scenario and identify the first signals that one of the four alternative worlds is producing. This observation could provide important clues in the business development phase, when the firm, concluding the reflection on the business model phase, enters into a more concrete and operational phase. Depending on the evolution of situations outside the control of the firm, managers can work with a view to fine-tuning the chosen business model in order to optimise the chances of success and reduce the firm's risk.

3.5
Enabling technologies

A central phase in defining a new business model is without doubt the identification of enabling technologies that the firm has at its disposal that allow arrival at a satisfactory solution to the initial problem. In this context, the term technology is understood in the broad sense. We are not only speaking of sophisticated engineering tools, covered by patents or trade secrets, but also implicit knowledge, relevant and perhaps unique competencies that the firm owns and is able to activate, possibly better than any competitor, in order to seize new opportunities. This knowledge is often found through a review of existing processes, also with the help of external consultants and tools such as those proposed by Merli in Chap. 4.

It should be borne in mind that owning and having sole control of knowledge/critical technology is not always a guarantee of success for the firm. This is certainly an important starting point, but not enough to obtain a new business model. Take, for example, the case of information technologies (IT): even their massive adoption does not necessarily lead to a proportional increase in competitiveness. In the decade that has just ended, several economists launched an intense debate to understand the real impact of IT on business processes and productivity in the services sector (see, for example, Carr 2004, as an interesting point of departure to explore the argument in question). Certainly the diffusion of these technologies is a necessary factor, but not sufficient to arrive at the realisation of those business practices that Rey speaks of in Chap. 7. What often happens in terms of the most significant knowledge that can actually become a source of competitive advantage is that the ability of appropriation and integration of business processes is not equally spread between firms and therefore only some of them are actually able to benefit from the "edge" that this privileged access represents. Going beyond the analysis of available knowledge and relevant technologies, identifying those competencies that we can manage better compared to competitors and successfully apply to solving individual problems, leads us to the true purpose of this phase of the business model innovation process.

The central question in identifying business models that can be turned into opportunities for commercial success is the following: of which competencies are we the real champions? In fact, successful business models are based on the exploitation of unique vantage points that are not easily transferable and the firm for some reason is able to exploit better than others. It does not necessarily have to be an absolute competitive advantage; it could simply be an element of differentiation that the company has developed from very specific contingent situations. This phase of reflection, therefore, brings business developers to think about the solution to the initial opportunity, perhaps not the one best suited to solve a problem but one that enables the firm to exploit certain positions of advantage, through a new product, a new service or a new product–service combination.

This requires verification of which previous investments made by the firm may be useful in enabling technologies that can exploit the opportunities identified through a new business model. These investments can be thought of as an iceberg floating in the ocean, emerging from the undifferentiated waters that surround it. For different markets, for different business models, some icebergs, namely certain factors of differentiation, are more relevant than others. What are the icebergs, technologies and points of differentiation that can be exploited to a greater extent in the current situation? An iceberg is composed of two parts. The submerged part represents the investment that a company has made as a precondition to entering into a particular technological domain. These investments are a necessary condition, but do not guarantee per se any capability to differentiate from the competition. However, these capabilities form the base on which a company can accumulate "differentiating knowledge", represented precisely by the tip of the iceberg.

The melting of the iceberg – but here we enter into a different discourse – illustrates the "commoditisation" trap, i.e., the gradual obsolescence of capabilities and core technologies that become progressively undifferentiated and part of the public domain. Hence, not only must the firm use these icebergs correctly, valorising them in their business model, but they must also continue to actively invest in maintaining this differentiated position. It is precisely the activation of new services that allows identification of new development opportunities (Chesbrough 2011).

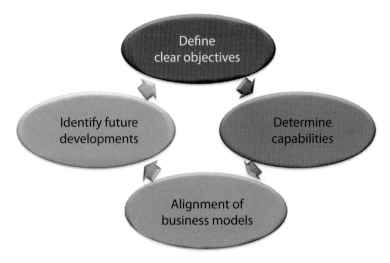

Fig. 3.5   The path to defining an alliance

3.6
Complementary assets and alliances

If the discussion on enabling technologies has been conducted with the goal of developing a new business, then, inevitably, at a certain point, an awareness will be manifested that not all ingredients needed to develop a new model are already available in-house. This therefore requires identification of the essential elements, currently external to the company, that will make it possible to construct a new business model. These are called complementary assets that the firm does not have but can be sought and obtained through supply agreements or strategic partnerships, perhaps aimed at realising a new innovative project.

This development phase of a new business model is particularly challenging and complex because often the risk of inappropriate alliances is underestimated or because the bargaining power a supplier has, at junctures where quality and compliance with the specifications of a component are fundamental, has not been fully considered.

The starting point of open innovation applied to the world of services, as in the manufacturing sector, is likely to be: "the most suitable person to implement the project does not work in this company, but his/her competencies are available on the market". The problem clearly concerns finding the competencies on the market needed to develop a new business model. Which alliances are needed to fully exploit the competitive advantage through a new business model? The management of alliances is likely to represent the riskiest option for the realisation of our model.

The success of the operation will at times depend on the behaviour of partners; the fact that empirical studies demonstrate that more than two-thirds of alliances do not achieve the objectives set is certainly not encouraging news. It would be particularly useful, especially for key development projects, to have models available that are able to eliminate the risk inherent in activating alliances, but this is clearly not feasible. However, it is possible to identify the critical points in the choice of partners and in the design of an alliance that are particularly important and that a firm must try to comply with (Fig. 3.5).

The proposed recipe is clearly not exhaustive but refers to a rather simple idea: alliances must be finalised and must have a clear purpose, namely to access complementary assets, and in order to function must be based on a clear and transparent logic.

Defining clear objectives. To establish an alliance the partners must have a clear goal. Why is this partnership necessary? Does it aim to improve margins? Or to reduce time to market? Increase the range of innovation opportunities? Expand the markets that the company can access? Defining the goals of an ideal partnership requires an outline of the specific characteristics of partners. It is not wise to embark on a partnership unless these goals are clear. In the course of developing the relationship, although this goal may change, starting out without objectives is extremely dangerous.

Determining capabilities. Once the objectives have been defined, a discussion needs to follow on what you are willing to share and what the partner is expected to put on the table. Indeed, it is not always possible to have the competencies available that the alliance requires. For organisational, strategic and confidentiality reasons or other contingencies, firms that embark on a partnership process must come to terms with what they are willing to put in the game. Certain capabilities are indispensable to the alliance function logic. If two firms are unable to make these resources available, then the alliance cannot work. Ancillary capabilities, instead, such as processes, activities and resources, if shared, could improve the results of the alliance. In this case, negotiation will define if and how to develop the relationship with the inclusion of ancillary lines of activities. Finally, there are those competencies that the partner firms consider critical for their businesses. These cannot be shared, on pain of losing autonomy or sharing information and strategic resources that the firm would do well to keep for itself. An examination of these key competencies is fundamental in undertaking a partnership, also to understand what our limitations are and where to draw the line that must not to be overstepped. Clearly, we may also be aware of competencies that are critical to our partners and that may become important to us. In this case, the only way to gain access to these competencies is by acquisition.

Business model alignment. Another golden rule in undertaking an alliance is to not consider oneself "the most cunning of all". Or at least, it is advisable to not behave as such. Translated into formal terms, in the definition of an alliance, the way in which to align the business models of the various partners must be identified, i.e., a second logic must exist according to which if everyone carries out their duties and contributes what is required – assuming that external circumstances coincide with what is foreseen – then everyone gains.

Otherwise, if this win–win situation does not exist, why would the partners remain in play? If a win–win solution is nowhere in sight, then it is likely that the real pay-offs of the partners are not those that were originally stated. Perhaps the alliance logic was not made thoroughly explicit and in this twilight zone pitfalls lurk, both for the success of the alliance and for the individual partners. Again, we must keep in mind the fact that once the appropriate objectives and competencies required from the ally have been defined, the search for effective alignment of the business models may turn out to be fruitless. In this case, the alliance will not work according to what was foreseen by the agreements.

However, even when business models are well aligned, this does not necessarily imply the absence of surprises in the development of the partnership. Indeed, one of the two partners in creating the win–win conditions could realise the objectives in a totally unexpected way and may perhaps be disproportionate to the remuneration requested for the commitment of the other partners. In these cases, a renegotiation phase is likely.

Identification of future developments. As noted earlier, a partnership relation is by definition risky and full of unexpected contingencies, reversals, previsions that are disregarded and models that are only half applied. It is therefore recommended that a partnership project be characterised by successive steps, milestones that mark some fundamental phases: a verification where those involved can decide to accelerate or modify the path, identify new targets. It is clear that knowledge and positive experience with a new partner progressively lead to increasing levels of trust, which can lead to opportunities for the development of a relationship that was initially un-contemplatable. If earlier, for example, some capabilities were considered critical and could thus not be shared, with time, these capabilities could be put back into play and provide the basis for the requalification of the relation and more ambitious projects.

3.7
The new solution: combining products and services

Once the available resources and complementary assets to be accessed have been defined, and with the initial problem clear in mind, the time has come to outline a possible configuration of a new offering to bring to market. Very often the object of commercialisation through a new business model will be the combination of a product with a service. On this theme, several contributions in this book may be helpful in addressing business development choices, starting with Dalli and Lanzara's Chap. 2 and Merli's Chap. 4. Some useful hints can also be found in Henry Chesbrough's (2011) book on *Open Service Innovation*. Here we would like to reflect on a key

Table 3.1 Expansion of the service offer (modified from Oliva and Kallenberg 2003)

	Services focused on products	Services focused on end user processes
Transaction-based services	*Basic services for users* Documentation management, transport, installation, basic training/updates, help desk, inspection, diagnosis, repairs/spare parts	*Professional services* Process-oriented engineering, process-oriented R&D, advanced training services
Relationship-based services	*Maintenance services* Preventive maintenance, monitoring, spare parts management, maintenance contracts	*Operational management* Managing maintenance function, managing outsourcing of functions

element in the product–service continuum, without repeating issues that have been clarified elsewhere.

More precisely, what relationship can exist between the base represented by a product and the services defined around it? In a business model outsourcing process, the percentage of the value-added component of "service" increases and therefore the percentage of the "product" component decreases. However, very often, the starting point is the customer of an existing product and it would hence be imprudent to disregard the characteristics of their demand.

Particularly helpful in this regard is the scheme proposed by Oliva and Kallenberg (2003) in their analysis of the transition of a product-oriented offer to a service-oriented offer. These authors identify the starting point in the installed product base, namely customers that are currently using our products, and identify two fundamental movements in the evolution of a business model in the services perspective. We use Table 3.1 to display this concept.

If we want to innovate starting from what we have available today, even when we want to introduce particularly novel, disruptive, unique and differentiating elements, we can attempt to ask two questions:

1) Are the services we are proposing directed at the use of a product or are they designed for integration with customer processes?
2) Are the services we are offering sold on a "transactional" basis or is the relationship with our customers "relational"?

From the intersection of these two dimensions we obtain the matrix shown in Table 3.1. Services that are oriented to the product on a transactional basis are base services that assist our installed base in the use of a particular product. Examples of these services are documentation and installation support, remote assistance and the supply of spare parts.

Services that are oriented to the product but that we nevertheless offer our customers on a relational basis are actual maintenance and monitoring services, and comprehensive management of the proper functioning of a given product, for example, complex machinery.

Returning to the transactional offer, but oriented to our customer's processes and therefore no longer to a single product, we here have activities that function from a professional services perspective. These are on-demand services that perhaps rely on instrumentation and products but that impact on, and often redesign, our customer's activities, for example, consultancy, training, testing and re-engineering activities. Finally, if this last activity is not undertaken on a transactional but on a relational basis, we are then taking responsibility for our customer's entire operations on an ongoing basis. We could for example arrive at entirely managing some of their operations or a particular need.

The move from a transactional to a relational offer and from focusing on a product to focusing on the customer's operations are two fundamental evolutions of a business model from an outsourcing perspective. This process must not however neglect the starting point of the firm, upon which its competitive advantage is based today. Clearly, not all models and not all activities lend themselves to these operations, but we must strive to find new ways and new solutions from the perspective of a progressive upgrade rather than a radical upheaval of the firm's identity. This concept applies to both the business-to-business and business-to-customer world. In fact, even in the latter case, some firms have been able to reorient their proposal based solely on the sale of a product, up to almost entirely taking charge of their customer's needs.

3.8
Value proposition

We thus come to a point in our path that requires synthesis of the new proposal, with clear and increasingly precise concepts, where we begin to see the risks and potential, but also the firm's bottom line.

It becomes essential at this stage to define a value proposition that is central to the definition of internally and externally relevant strategic documents. From an internal point of view, the value proposition has the purpose of preparing the ground for the discussion on the investment opportunity and hence, from a business development perspective, the moment will come when the project management team must prepare a presentation to top management in order to obtain their go-ahead for the operation. This step is particularly necessary to explain the feasibility of the operation, consistent with the values and mission of the organisation, as well as congruence with the expected ROI.

From an external point of view, the value proposition will constitute the preliminary core that will help define the marketing plan describing the value added for the final customer.

It is useful at this stage of the process to open the tasks of the business development groups to greater sharing with the rest of the firm and eventually with external parties that may be able to contribute to this phase of the business model innovation process, which on one hand should be highly creative, but also very realistic and in line with the firm's values and opportunities. In fact, on presenting a new message to the customer, we must be satisfied with the effect that this message will have, but we must not forget its consistency with the content thus far transmitted or, and above all, the risk of expending the firm's name on unrealistic promises.

The key to constructing a new business model is therefore to arrive at, and effectively synthesise from a communication point of view, a business idea in line with corporate values ??and realistic in its possible implementation. However, arriving at the synthesis of a creative process is not simple. Perhaps there are numerous ideas/visions, some of which are complementary, others in sharp contrast, and this fervour is the result of a creative process that has worked well. An exercise that can lead to the necessary synthesis of a value proposition is to build what the Americans call the "elevator's pitch", that is, a very brief presentation that explains in a few essential elements the concepts of the new business model. The pitch consists of a logical and very linear sequence that should include references to:

1) *Target market*: For whom is this new product/service intended? This is about a market segment that is particularly sensitive to our offer. In a second phase other markets could develop, but for the initial launch phase this category will be interested in our offer first.

2) *Characteristics of the target*: Which of the many features of this market segment are particularly interesting for our offer?

3) *Identification of the offer*: How do we characterise our offer? Here it is important to start from a known product category: to be understood, our offer must seek resemblance of some characteristics of its use with something that already exists.

4) *Key benefits*: Our offer, however, is distinct from the product/service identified because it does something that others cannot do.

5) *Comparison*: In a comparative perspective, the pitch explains the benefit of our offer by comparing it with the value generated by other offerings in the marketplace.

6) *An element of differentiation*: For each offer on the market the pitch explains the substantial difference that qualifies our key benefits in the customers' eyes.

If thinking, for instance, of inventing the concept of fast food, our service will primarily be appreciated by a very young audience (target market), who want cheap, tasty and delicious food (characteristic of the target). We will perhaps have a chain of owned or franchised eateries that distribute a modular menu consisting mainly of a side dish (fries or salad), a hamburger with a range of alternative dressings and a soft drink (identification of the offer). This menu consists of pre-cooked items that can be assembled and delivered in less than a minute from ordering and costs a maximum of €10 (key benefit).

Unlike a traditional catering service or a quick sandwich at the bar (comparison), our product can be taken away and is quick to eat but remains a complete menu; it proposes to be the most economical and delicious hot meal on the market (differentiation).

It is clear that these concepts – expressed or represented in a series of slides with a video or suchlike – will pave the way for a discussion and give rise to requests for additional information. But the goal of the value proposition is not so much exhaustiveness as clarity and persuasion, to then proceed with specific insights, perhaps concerning an investment plan, a marketing plan or, more generally, the next steps to implement in the new business model.

3.9
Conclusions

At this point, every team should have collected enough information to complete the structure of the business model in Fig. 3.3. This final synthesis makes it possible to effectively communicate the work carried out and the new end point in the product/service combination that best lends itself to responding to the initial problem. Completing the business model canvas is a useful exercise to clarify ideas on the results obtained and also to have to hand a rather effective illustrative tool.

It is clear that this end point could also represent the starting point of a new path. Business models, in fact, do not respond to mathematical laws and their validity is based on assumptions that are nevertheless risky and depictions of scenarios that are far from complete. A "plan B" has to be ready when the assumptions on which the reasoning of starting the new business model rested begin to creak. In the analysis of several new businesses, Mullins and Komisar (2009) confirm that it is not so much on the first, but on the second or third attempt, that the firm achieves the hoped-for success.

Thus, we close this chapter by reiterating that, in the service sector, business model innovation requires a business entity that bears the risk inherent in any innovative project. It is the task of analysts to attempt to find the logic that minimises the level of risk of a new business venture and it is to this end that this chapter presents a framework to guide the innovation path.

References

Carr NG (2004) Does IT matter? Harvard Business School Publishing Corporation, Boston, MA

Chesbrough H (2003) Open innovation. Harvard Business Review Press, Cambridge, MA

Chesbrough H (2004) Managing open innovation: chess and poker. Res Technol Manag 47(Jan–Feb):13–16

Chesbrough H (2008) Open: Modelli di business per l'innovazione. Egea, Milan

Chesbrough H (2011) Open service sciences [in Italian]. Springer Italia, Milan

Garvin DA, Levesque LC (2006) A note on scenario planning. Harvard Business School 9-306-003. Revised 31 July 2006

Mullins J, Komisar R (2009) Getting to plan b: breaking through to a better business model. Harvard Business Press, Boston, MA

Oliva R, Kallenberg R (2003) Managing the transition from products to services. Int J Serv Ind Manag 14(2):160–172

Osterwalder A (2010) Business model generation: a handbook for visionaries, game changers, and challengers. Wiley, Hoboken, NJ

MAINS Master, academic year 2007/2008
People and companies involved in the InnoLab:
Students: Filippo Barra, Roberta Ghedini, Francesco Inguscio, Donato Mazzeo and Alice Orlich
Companies: Centro Ricerche Fiat, Elsag Datamat, IBM Italia, Tiscali and Xerox
Professors: Riccardo Giannetti and Paola Miolo Vitali

1. The problem
The workshop focused on the emerging opportunities in the infomobility services sector with the aim of identifying an innovative service and analysing the relevant business model.

Infomobility as an area of research and technological development can be defined as the use of intelligent transportation systems (ITS) technologies to improve the management of public and private mobility, reduce congestion and environmental impact and therefore improve the quality of the life of citizens.

Innovative services that can be accomplished in this arena have a significant effect in economic and social terms. The inherent complexity in identifying infomobility services and related business models lies in the correct identification of the real needs of users and in the correct analysis of the ecosystems that such services can enable. The interaction between the different actors of these ecosystems must be analysed, defining new business processes and the competencies necessary to achieve them. The service and the relative business model proposed by the lab, valorising the core competencies of partner companies, would have to leverage on an open ecosystem, putting into practice the "open innovation" philosophy theorised by Henry Chesbrough.

To analyse the business ecosystem at the base of the services studied, Component Business Modelling (CBM) – introduced by IBM – was the lab's chosen methodology. CBM was established as a framework for mapping business components, recombining them in order to concentrate firm resources on those with greater value-added, thus maximizing the value maximising the value created by the firm. The final objective was to complete the business model with a proposal for a business case able to highlight the economic sustainability of the solution identified.

2. Work methodology
The work was carried out in two main phases:

1) Mapping the full range of infomobility services.
2) Identifying a specific area to be analysed with the CBM methodology and proposing within this an innovative service from which a business model and relative business case could be outlined.

In the first phase, the most important relationships in the industry were studied while the mapping effort allowed identification of the macrofamily of services together with a description of the main categories of domestic services within these. We thus identified five categories: (1) individual transportation management, (2) freight management, (3) traffic and safety management, (4) mobility payments and (5) advanced vehicle control.

The analysis of each category was implemented by taking into account the following variables: scope and functionality of each service, the relative needs to be met, technologies, adoption rate, current trends, major benefits for the firm adopting the technology, adoption restraints, ecosystem actors and case histories.

The mapping led to the decision of analysing parking payment systems. Studies on the ecosystems of businesses related to parking payment systems are neither numerous nor extensive. The nature and purpose of the research thus led to the use of a methodology based on exploratory case studies. It was decided to focus the study on the Municipality of Pisa's payment management case history, analysing empirical data related to the different methods of payment adopted.

The case study examined interviews with the mobility service company PISAMO SpA in Pisa, conducting an analysis of the service list and company documents, as well as additional documentation (various presentations, etc.). Various payment methods were identified around which an analysis was subsequently structured: (1) cash, (2) subscriptions, (3) Europark electronic card, (4) scratch and park and (5) mobile phone payment system.

The business processes were then mapped in relation to the functioning of each method and the Component Business Model was subsequently outlined on the reality of Pisa's parking payments, attempting to generalise the analysis on the basis of available documentation. In this context, CBM was used to identify the business components by mapping the ecosystem in which the firms interact.

Amongst the main actors involved in the ecosystem, the mobility services firm was identified as having a central role, alongside several complementary players: parking meter suppliers, the traffic warden firm that controls traffic and infringements of the Highway Code, the bank (especially involved in cash counting activity), as well as several other actors with secondary roles. For each actor the relevant business components were identified, highlighting key inefficiencies by method and actor of reference.

In parallel, scouting was undertaken of the most promising technological innovations in the mobility payments industry using the strategic and technological competencies of partner firms.

Through a detailed analysis of the expressed and/or implicit consumer and business needs of users of the parking payment system – motorists on one hand and the urban mobility management firm on the other – an evaluation grid was constructed to classify the five methods according to the value created for users. The grid was also used to identify, among the possible payment methods based on innovative technologies, the method that brings the most benefits to the public and to the firms. In doing so, the lab took a "demand pull" approach to the introduction of the innovation.

3. Proposed solution

The solution proposed as an innovative parking payment service was based on the installation of traffic sensors in parking zones. The concept was called "iPark" in order to highlight the "intelligence" incorporated in the sensors to determine occupancy of the parking zone. In fact, the roadside sensors in question were designed to communicate with sensors to be installed on board vehicles, activating an automatic parking payment process.

From a marketing point of view, iPark is able to match and exceed the benefits that citizens and mobility management firms achieve through other payment methods. Furthermore, additional benefits were identified following the adoption of this service: the system would enable reservation and payment in real time for motorists and favour traffic decongestion management, allowing the mobility firm to dynamically modulate parking fees.

As a consequence of the CBM construction, the main novelty that emerged in the business ecosystem was a new actor called the "sensor management firm", in charge of the implementation and management of the sensor system as well as the entire infrastructure and the data produced in real time by the system. The eventual introduction of new activities for actors already in the ecosystem and the greater efficiency of processes implemented by them thanks to the introduction of this new method should also be highlighted. A significant example in this regard is the improvements that could be made to the process of detecting infringements thanks to "targeted" alerts.

Among the main benefits emerging from the mapping of business processes was the possibility of enabling new services for users through the iPark infrastructure (e.g., displaying occupancy data in real time as content for satellite navigation firms, etc.).

The model that was adopted to test the economic viability of the service assumes that the investment in infrastructure is made by the sensor management firm and the use of the infrastructure is offered as a service to municipal urban mobility service companies. The business case was conducted using conservative assumptions on both the number of units with sensors and in terms of user migration to the new payment method.

Through the quantification of benefits such as reduced costs and increased revenues associated with the migration from old payment methods to iPark, it was possible to demonstrate the benefits of the investment for the local mobility firm from year one. In the case of the sensor management company, the same assumptions demonstrate the possibility of achieving break-even within four years and a high return on investment in subsequent years.

The transformation of the business model: business modelling

4

Giorgio Merli

The reconfiguration of the business model of an enterprise, especially to enable new businesses capabilities and/or operate in an "open business" logic, requires organisational know-how and tools that are a little different from those traditionally used. The representation of the business necessary for this purpose is the mapping of activities/competencies that the firm needs in order to participate in the business ecosystem in which it operates. This helps to understand what activities are needed to create new value propositions and/or to ensure competitive advantage differentiation and/or to evaluate the possibility of outsourcing. This logic is nevertheless crucial to be able to constantly realign the organisation on the segments of the value chain that have the highest value added or are more protectable/unique, at the same time identifying opportunities to reduce structural costs. This "mapping" of the business is in fact the firm's business model. IBM has developed a specific methodology in this respect called Component Business Modelling (CBM).

4.1
The need to transform the firm's business model

In the current evolutionary scenario, the transformation of the firm's business model is a mandatory[1] step, which makes it possible to:

1) enable those new business capabilities identified as necessary for a new business strategy (e.g., to give shape to the most innovative value propositions or new service activities);

G. Merli (✉)
Dipartimento di Sociologia, Università degli Studi di Milano-Bicocca, Milano, Italy
e-mail: g.merli@yahoo.it

[1] On this point see Chaps. 1 and 3.

L. Cinquini, A. Di Minin, R. Varaldo (eds.), *New Business Models and Value Creation: A Service Science Perspective.* Sxi 8, DOI 10.1007/978-88-470-2838-8_4, © Springer-Verlag Italia 2013

2) activate new paths of value (i.e., new ways to generate revenues and margins, or new positions along the value chain);
3) create a more flexible and streamlined organisation and/or reduce fixed business costs by freeing financial resources (very important in situations of financial shortage).

Enabling new business capabilities means building new strategic, operational and management capabilities, to give shape to the new value proposition and/or new ways to propose or take them to market. This evolution may be a result of the need to align with the competition or, rather, to forestall the competition on innovative offerings. This would allow the creation of a situation of "Surpetition", namely, a decisive competitive advantage that can avoid competing with prices on the same value proposition as competitors. In the case of a strategic decision to "servitise" a product/service that already exists in a traditional manufacturing firm (see Chap. 2), the first step is usually to create a separate organisational entity with specific competencies. The existing one has to be maintained until the new competencies become pervasive within the enterprise and the offerings are fully integrated and completely servitised. This is what occurred in the automotive, or generally in the machinery, industry: at the start of the evolutionary path towards services, companies created real and virtually independent business units to offer, for example, financial services or the sale of managed fleets.

The offerings were gradually and subsequently integrated into a single form of sale of the servitised product and the new business capabilities became inseparable from the previous ones. New competencies, new capabilities or new activities could of course be activated within the boundaries of the firm or with partners or suppliers. However, this is a modification (often enrichment/extension) of the firm's business model. The impact on business and support processes is without doubt significant.

Creating new paths of value means activating new business segments in the face of new value propositions as mentioned, running new "rings" in the value chain in which the firm participates (for example by moving downstream towards the customer or upstream towards the supplier) or, finally, occupying business value-added "nodes" in the ecosystem in which it operates (e.g., make or acquire a new technology to serve another value chain of businesses that is in some way "related"). Manufacturing firms are in this way selling maintenance services also for the competition's machinery. Services designed to support the business become autonomous and independent business entities and new sources of revenue and profit for firms. Financial services companies are often born within large groups to provide financial services to the companies of the group and then in many cases evolve and deliver financial services to the external market. Zara integrates competencies and capabilities of fabric and leather processing internally, buying the product in advance in order to be able to react quickly to consumer taste experiments carried out in-store.

Traditional manufacturing firms integrate downstream, becoming actual retailers and creating an end market for themselves. Giovanni Rana created a chain of restaurants to distribute quality Italian food around the world and is also involved in initiatives in America to promote the Italian food supply ecosystem. Retailer chains are opening service stations, occupying a "node" that was previously ungarrisoned in the ecosystem of the relation with the end consumer. In addition, the firm's business model must be reconsidered in the terms previously mentioned. Innovating the role/position in the firm's value chain means knowing how to concentrate/focus on those segments of the value chain where it has greater competitive/distinctive capabilities (such as engineering or manufacturing) and/or those that are vital for its total governance (e.g., marketing/dealership or the concept/design of the product if the brand is strong), or simply the most easily defendable. On the other hand, it may simply mean the ability to modify its supply chain, possibly even involving competitors with different configurations for each product/service, to achieve maximum effectiveness and efficiency (by now standard, for example, in engineering and car manufacturing).

Knowing how to create leaner, more flexible and responsive organisations and reduce fixed costs while freeing up financial resources is crucial in situations of economic and financial crisis and in all short-term difficult market situations. This approach essentially means knowing how to reduce the number of activities carried out internally by the firm, outsourcing all those that do not directly contribute to what is perceived as value by the customer (previously termed "added value for the customer") or are not factors of competitive advantage for the business. Obviously, this does not mean eliminating them, but letting them be managed by other companies, perhaps specialised in such areas. All this must clearly be combined with the simultaneous development/improvement of those assets that instead provide firm competitive advantage/differentiation, to avoid a negative financial turn of events (that only continues reducing internal activities and hence, sooner or later, depletes competitive capabilities). The concept that underlies this logic is that it is (clearly) worth investing financial resources in activities in which the firm "is good and wins" rather than in those that "do not make a difference", keeping financial resources tied up in low-leverage business. The need to review the business model to this effect is often combined with what follows herewith.

Creating new and more flexible business organisations may mean to realise a dramatic change in the firm's organisational strategy, pursuing the capacity to create/reconfigure the organisation as a function of opportunities or threats or business problems that the scenario presents. Knowing how to react ahead of competitors, perhaps in a proactive way, can constitute a decisive competitive advantage. This strategic organisational capacity will obviously have to be met with dynamic operational management processes that are also able to continuously reconfigure the function of the specific business/opportunity, including considering all sourcing options (outsourcing, insourcing, co-sourcing, near-shoring, offshoring, etc.). This strategy of rendering the business model more flexible can coexist with all the remodelling mentioned above, creating a competitive advantage on a continuously winning transversal dimension, namely, speed/flexibility. This is in fact a general requirement, considering that business is progressively executed in increasingly large and open business systems (business ecosystems–open business), with continuous strategic repositioning opportunities and the use of partners to rapidly develop new businesses.

4.2
Focusing on core activities

Many of the issues cited on the motivations and methods of reviewing the firm's business model recapture the importance of being able to concentrate and develop those activities that provide, or will provide, a clear competitive advantage.

By contrast, those other activities that do not provide effective competitive advantage, and thus cannot qualify the firm's success, need to be very clearly identified as well. The fact that they may be necessary or indispensable for the realisation of the firm's supply chain or value chain does not mean they are factors of competitive advantage. Transportation of a car or invoicing are not distinguishing factors for success in the marketplace. Its design certainly, its production…perhaps. The desire to use a strategy focused on internal core activities, with simultaneous recourse to specialist third parties for non-core activities, can be defined as a *specialisation strategy*. This strategy assumes that the ideal would be for a firm to specialise in and concentrate on those skills and/or activities where it manages to achieve the best competitive advantage/differentiation and should instead make recourse to "others" (suppliers and/or partners) to carry out the other activities necessary for its business.

The possibility to activate a specialisation strategy is today much greater than in the past. This is due to the fact that business ecosystems are improving around the world and especially because almost all contributions that are considered non-core are much more easily accessible (thanks to computers, the internet and technology in general). Even transaction costs are greatly reduced and in many cases are now close to zero. Barriers of "time" and "distance" have become negligible for many activities. The world has become much smaller. Operations and financials are more visible and the risks of collaboration have been considerably reduced. It is now much easier to find the best practices and use them, even if physically distant.

In conclusion, in order to activate this strategy, a firm must focus on its "specialisations", where it is able to express its competitive advantage ("internal" specialisation) and make use of the contribution of a network of "external" specialists (probably from the same industry) to achieve the responsiveness/flexibility and the efficiency that can make its supply chain more competitive. Finally, it should pool all its support activities (HR, IT, etc.) to create greater economies of scale. These activities can still be managed internally by creating Shared Service Centres, co-sourcing with others (even competitors) or outsourcing (possibly near-shoring or off-shoring), or even in-sourcing, if able to make such services attractive to others (in this case, they must be managed as profit centres or companies in their own right). Strongly prevailing is the solution of having them managed by external specialists (usually "transversal" partners/suppliers, not specific to the industry) that can achieve significant economies of scale useful in reducing the costs of these "undifferentiated" activities (they can count on higher volumes, bringing together the activities of multiple customers). Into this category fall the activities of managing payrolls, benefits, logistics (both incoming and outgoing, and perhaps even internal transportation), IT, call centres, technical assistance, etc. This strategy can furthermore achieve the simultaneous goals of increased competitiveness, reduction of fixed costs and increased structural flexibility, as advocated by previous strategic arguments.

4.3
Continuous reconfiguration capacity

The capacity of reconfiguring the firm's business model requires organisational know-how and tools that are somewhat different from those we have used up to now to implement traditional corporate reorganisation projects. Indeed, it would be difficult to deal with such a complex and delicate subject by simply reasoning on organisation charts, functions or cost centres. These, in fact, misinterpret the logic and granularity that are needed to determine, for example, "what capabilities/activities enable differentiation and/or offer competitive advantage" or "which activities could be outsourced". However, this would be just as difficult to undertake even when using the most sophisticated know-how of process organisation. The processes and their sub-sequences are in fact value-added activities aimed at the production of some output. Know-how developed in this regard is designed to garrison, govern, manage and improve the flow of activities necessary for their operation. If we take, for example, an engineering process, this represents all the activities that have to translate a project into *bills of activities and materials*. In today's realities, this hardly coincides with activities entirely implemented within an organisation, given the continuous recourse to external inputs and the correlated ingoing–outgoing information. In these situations it would be very difficult to get an entrepreneur/CEO to talk to an organiser and a planner of what could be outsourced on the basis of the engineering processes/subprocesses. It would be much more appropriate to talk about the whole engineering activity, or about a subset of it, or about designing competencies/capabilities, or better yet about the *body* that could engineer what we are discussing. This body is neither a traditional function nor a process/subprocess (even if able to deliver it), but more properly an "operational entity able to perform a complete set of engineering activities/services with a concrete perceivable value in the market". According to this logic, examples of operational entities on which to assess whether, for example, "we are able to perform it" and/or we can outsource or acquire it, are "legal office" or "outbound logistics" or an entire factory. The logic and granularity of these entities are therefore aimed primarily at the questions: "Is it strategic to my competitive advantage?" and "Do I keep it or outsource it?".

The representation logic of business activities that serves this purpose is hence the mapping of activities/competencies in terms of "competitive advantage/differentiation" and their possible outsourcing. This is crucial to be able to conceptually and continuously realign the business with the paths, or rather with the segments, of the value chain with higher value added or that are more protectable/unique.

However, it is also important to quickly identify the activities that can be out-sourced to reduce or variabilise fixed structural costs without affecting the firm's competitiveness. A "mapping" of business activities with this logic can be identified as a business activities or components model.

4.4
The business activities model

A methodology developed by IBM to address the issue of business modelling useful for the purposes described in the preceding section is Component Business Mod-elling (CBM). A business component is defined as a homogeneous group of busi-ness activities, equipped with their own information system, processes, organisa-tion, governance and performance indicators, capable of generating specific added value to the business enterprise. These components are characterised by the type of contribution to the business (core, non-core, competitive levers, support, etc.), also in a time-varying way. The search for the "necessary and sufficient granularity" to identify the primary organisational element can be identified in the principle of "its transferability to another organisation without losing the functionality of the rest of the model". Examples of this could be the "legal" or "communication" function, or part of the activities performed in these functions, but also a function that is more closely linked to the business, such as "purchasing". These activities can be consid-ered "components" in as much as they can be treated and independently outsourced in "services" without any particular constraint, if there are no business contraindi-cations.

In a "processes" model, these could not in fact be identified as a process (since generally they are simply not). In a "processes" logic, these are considered activi-ties that are functional to processes such as "litigation management", "communica-tion processes" and "procurement". Processes, in fact, incorporate or involve sev-eral functions/activities. The procurement process, for example, is typically achieved with the contribution of the following functions/activities: product development, en-gineering/technology, product and production planning, quality, logistic/entry in-spection, accounts payable, etc.

This last example enables focusing better on the two main differences be-tween process/subprocess components and business components: "granularity" and "nature".

Different "granularity". Generally a business component (for example, "opera-tional purchasing management" activities) is smaller than the processes that "use" them (such as the overall procurement process, from strategy to procurement plans, purchasing, delivery, incoming inspection, stocking or delivery to production lines), but nevertheless has a well identified added value and can be independently managed.

But it can be even much larger, such us a whole factory. The manageability as a self-standing set of activities identifies what can be considered a "Business Component".

Different "nature". The process always foresees activities that are "concatenated" along a value chain or supply chain, while the component can also be an activity/function (such as procurement) in "service" (also *on–off*) to some business components operating on the value chain as a core process/activity.

4.5
The structure of the component business model (CBM)

The CBM model organises all the components of the firm according to two dimensions:

1) the level of "management" of the component;
2) the area of activity.

The level of management of the component identifies the type of activity to be performed and the level of "responsibility" of the activity. It can thus be identified as a strategic type activity, a management and control type activity, or an operational/execution type activity. The areas of activity indicate the nature of the activities and groups together homogeneous activities by competencies and resources. Areas of activities are related to the type of business and industry in which the firm operates. Areas of activities typical in the industry of consumer packaged goods are, for example, product development/management, market/customer, supply chain, manufacturing, business administration and support (which includes all typical support activities such as finance and control, information technology management, human resource management, etc.). Each industry is characterised by specific activities. In the energy and utilities sector, for instance, areas of activities are identified in network management and maintenance, energy production, energy distribution, relationships with network operators and the role of the firm in the overall ecosystem (producer, distributor, retailer, etc.). The model is detailed to an appropriate level to be representative of the firm's realities. Within the model, the activities that are specific to the sector are then identified. In banking, for example, there are models for retail banking and others for private banking, and the activities are detailed and specific to the sector such as risk and financial management. Starting from the industry models, a specific map of the individual firm is drawn up; the management level of the component is recorded in the rows and the areas of activity in the columns, as shown in Fig. 4.1. Combining the two dimensions (management level and area of activity), the model identifies for each area of activity the components that are required for the various management levels. The management/strategic type components are therefore components that include the activities of setting the guidelines and strategies needed to manage the different areas of the business.

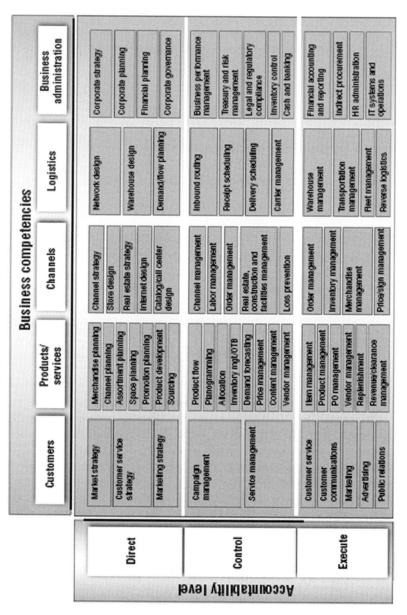

Fig. 4.1 An example of business model mapping in a CBM logic (company operating in the retail sector) (http://www-935.ibm.com/services/us/imc/pdf/g510-6163-component-business-models.pdf)

Some examples of this are market and channel strategy, industrial operations strategy and corporate governance strategy. Management and control type components, as the term itself indicates, include planning, management and control type activities. These activities are characterised by a single area of business. Some examples are planning and monitoring marketing activities (campaigns, promotions, etc.) in the commercial area of the different sectors, procurement planning and control, production planning and control, planning and management control of assets and the energy distribution network, and planning and control of claims in insurance companies. Execution type components include operational activities. Activities that may be of a production type in a manufacturing and industrial context are operational execution of marketing campaigns, operational management of a call centre's inbound and outbound calls in the telephone operator's customer relations area, the operational management of the firm's informatics infrastructure or facilities, operational management of a telephone operator's or energy distributor's network maintenance and operational management of the payment of premiums in an insurance company. It must be emphasised that the columns do not represent organisational functions but may in many cases coincide. The strategic evolution of the business model towards greater overall flexibility and responsiveness can lead to the correspondence of the organisational component and the entity when the component actually becomes a set of activities that are entirely "transferable". With these specifications it is clear that the representation of the firm's business in a CBM logic essentially responds to a very pragmatic logic of modularity designed to easily manage/reallocate activities. This is based on an articulated construction into business components (groups of homogeneous activities) that can be quickly/easily transferred outside or brought back inside or aggregated to others (internally or externally), depending on the strategic, or even tactical, priorities at the time. This constitution thus allows continuous reshaping of the firm's boundaries (an "amoeba" firm), garrisoning internally what is considered as most important at the time in order to be successful on the market (and/or to better manage operational costs) and outsourcing the rest. The desired "flexible configurations" can thus be created. This enables conceptual "real-time" management of decisions on outsourcing, insourcing, co-sourcing, offshoring and near shoring of business components. As previously mentioned, with this approach the components have a different granularity compared to the possible segmentation by process/subprocess, inasmuch as they automatically seek the defined and functioning elementary unit that is independently "manageable", and not a process or subprocess as such. We reiterate that this granularity is generally lower than in those processes usually considered, but this is not the norm. In fact, the reverse, albeit rarely, can also be true. Indeed, the defined operational entity that is independently manageable may inherently or pragmatically (for convenience or managerial homogeneity) span multiple processes. It may, for example, be more convenient/easier, in terms of outsourcing/insourcing decisions, to consider all HR activities (payroll, personnel administration, etc.) as a single component to be managed in an integrated way than considering the componential complexity associated with their greater granularity. The CBM methodology is particularly useful when designing a new business model that is required for new business configurations. This refs to a "to be" business model compared to an "as is" business model to identify the different activities/performances to be bridged and thus outline the necessary business transformation plan. Governance of the organisation in the CBM logic becomes increasingly necessary the more the firm has to face a continuous need for change. However, it is also methodologically very useful for a one-shot firm reorganisation project. Bruce Wright, senior vice-president of Bank of America Cards Services, made an interesting affirmation: "CBM is a radical new approach, inherently adaptable, able to cope with the ever-changing scenario of business […]. A powerful tool to link business and technology […]. A methodological approach that enables reaching decisions quickly to […]. Enables starting off small, even with a single component, obtaining quick successes that are useful to decide on the most important transformations".

In terms of the articulation and characteristics of the operative governance and responsibility of business components, these usually concern wider and more "entrepreneurial" roles than traditional roles (these are, in fact, "autonomous business components").

4.6
Planning and managing the transformation of the business model

As previously argued, the capability to understand and reconfigure the business model to best interpret the changes in the external environment is an important competitive advantage for an enterprise. However, the transformation of the business model has to be carefully implemented to enable simultaneous pursual of short-, medium- and long-term objectives.

The review of the business model should in reality be a continuous activity, with constant refinements and extensive breakthroughs. It is therefore important to operate a lean and effective methodology to manage this process. Let us investigate.

The *first methodological step* to implement a transformation of the business model is to create a map of the firm's activities according to the CBM model, or rather, "unpack" the firm into its fundamental components. Typically, the total number of components of a firm to be considered for this purpose varies from 50 to 70. Pushing beyond the granularity would not be methodologically and strategically significant since this would lead to an analysis that would be ineffective in terms of the objectives and even impede a strategic evaluation of the firm's business model.

The *second step* involves the identification of the key components. If new activities are to be activated (such as service activities or "servitising" existing products), new business components must be added to those already existing and mapped. In fact, in this case the current business model may not yet include the necessary internal activities.

Generally, therefore, four types of components need to be identified:

1) the new components required;
2) the differentiating components;

3) the components that enable a flexible and responsive model;
4) the strategic and operational components that enable the strategic goals in the short-, medium- and long-term to be achieved.

The *third step* is to assess the current capability of firm resources (competencies, skills, organisation and energy) and its renewability potential.

The *fourth step* is the synthesis of the assessments made previously. What emerges from the intersection of evaluations made, in the comparison between "where we want to go" and "what the actual/present level of resources is", is the configuration to pursue (the "to be" CBM). Thus, the necessary steps are identified to achieve this competitive configuration and the roadmap for the relevant transformation is outlined.

We illustrate some of the elements in relation to how to proceed in identifying/evaluating the four types of components. Four specific questions need to be answered:

1) Do we want to create new value propositions? If yes, what new activities do we need?
2) Which components differentiate/will differentiate the firm in the market compared to competitors?
3) Which components provide/will provide the firm with the capacity to easily reconfigure itself to the new structure (flexibility factors)?
4) Which components/activities will most enable the achievement of the business objectives?

The *first question* is closely linked to the firm's decisions on strategies to revise the value proposition (if there are no changes you move on to the second question).

The *second question* addresses the areas in which the firm has to be strong, structured, visible internally and on the market. This approach foresees identifying differentiating factors and evaluating the contribution of various components to innovation. If the differentiating factor is the customer relations capability in all its forms then the components must be identified that enable this factor. If the differentiating factor is the innovation capability, all the components feeding it should be identified and highlighted. The "differentiating" components can be of a strategic/management, planning and control and/or operational type.

Indeed, even some operational components can be differentiating levers. For example, for an online reservations company, one of the differentiating components of the model could be the call centre's customer management operational activities. The evaluation is primarily of a qualitative type and must still be undertaken in a "business" logic (i.e., with intuition rather than scientific/analytic). It can nevertheless be made quantitatively using a scale of values, for example from 1 to 4, with 1 being a component whose contribution towards the desired direction is low or nonexistent, while 4 indicates a component that contributes to the maximum. The evaluation could be even more sophisticated and objective by identifying the relative percentage weights of each element. In this way, the assessment of the components becomes more and more modulated and immediately distinguishes the relative contribution of each component towards the ultimate goal.

The assessment can be made directly by top management or by a team of first-level managers or even jointly with a talented team from different areas of the company. External support could be of particular help when the company needs an "external view" or otherwise for "impartial" methodological guidance.[2]

The choice of the approach depends on the type of objectives, the characteristics of the firm, its circumstances, the time available and, most importantly, its culture. If the firm is in a time of change, is significantly "shedding its skin" following a strong push by top management, then often an assessment of the components made directly by top management can be effective, possibly supported by a very small core of trusted collaborators. When the situation is less critical, where the corporate culture and the firm's situation allow greater sharing and more time, a path involving numerous actors of the firm on different hierarchical levels and from different business areas can also be very effective. For example, two working groups could be created: one composed of top managers and one composed of talented people from different business functions. The evaluation of differentiating components is assigned in parallel and the results compared in a joint session. This method presupposes firm and top management openness to discussion and an open dialogue. There are other options. The best path, as mentioned, has to be chosen based on the firm's culture, characteristics and situation, and the time available and the objectives. The result of this activity is an assessment of which elements are truly differentiating for the firm, i.e., those in which we need to invest to strengthen and develop competitive advantage.

The answer to the *third question* requires identification of which components are most important for a flexible and responsive business model. The evaluation approach is similar to that just described, but obviously the parameters to be considered for evaluating the components change. Flexibility and responsiveness are connoted by the context in which the firm is moving and should be deployed into some element that allows an accurate evaluation. Flexibility may be sought, for instance, in production, with scalable production models, or in the sales force, with resources balanced in fixed and variable components, in the capacity to convey the product to market and in the capacity to perform an administrative process. Clearly, everything is linked to the type of business and its characteristics. To identify areas requiring business model flexibility requires starting from the identification of the factors characterising the flexibility and responsiveness of the firm's ecosystem. Thereafter, the key flexibility and responsiveness factors need to be identified. Typically, there are 5–6 factors enabling the firm's capacity to respond quickly to the changing environment and the level of permeability with the outside world. These factors are declinated in various items in various business areas, such as in the capacity to quickly introduce a new product/service to market and/or in the capacity to seize new business opportunities or in the level of external diffusion. It is also important to arrive at a definition of what components are needed to "renew" the capacity of the model to be flexible within today's context, but also in future scenarios.

---

[2] To elaborate the final objective, the brainstorming proposed by Chesbrough, Di Minin and Piccaluga in Chap. 3 of this book may be particularly useful.

The answer to the *fourth question* brings the "objectives" dimension into the analysis. The objectives are clearly linked to the firm's situation, its contingencies, and the short- and medium-term horizons. Objectives may be of a financial, economic, market, environmental or social order. The "lenses" with which to analyse the firm's business model should therefore be identified according to the priorities. A few examples follow. If the company aims to reduce costs, the priority "lens" to use is the absorption of costs by component. If the objective is to reduce working capital, the lens to use is the identification of those components that contribute most to generating stocks/WIP, payment delays, etc. If the objective is to improve energy efficiency, the analysis should focus on identifying which components contribute most to achieving this goal. In general, the starting point of this analysis is the first deployment level of strategies and objectives of competitiveness and financial performance of the firm. A method widely used is to take as reference the firm's critical success factors at that particular time. Some may relate to competitive capabilities (time to market or innovation), others to operating performance (operating costs, efficiencies, sales volumes, etc.), others to financial aspects (working capital, etc.) and others to aspects of corporate social responsibility (carbon dioxide emissions, etc.).

Once the evaluation on three fundamental aspects has been consolidated, we must define the direction in which the model is to go to implement the transformation. It is thus necessary to assess the current state of performance of the business model to understand which transformation initiatives must be implemented. This type of analysis can be more or less quantitative in relation to different variables:

1) type of information;
2) time;
3) the objectives of the analysis.

To be able to implement an "entrepreneurial" type evaluation, the analytical level must be appropriate for the purpose. If is excessive it will require too much energy and take too long. If it is too approximate it could lead to incorrect indications. As is evident, building upon existing information is expedient. If, for example, a firm is already structured with an activity-based costing model then it is easy to allocate the assets and costs in the components map previously drawn up. If this is not the case, and this happens for most companies, a very effective way to proceed is through reasonable assumptions in order to relocate the activities and related costs. The evaluation can also be done qualitatively, from a relative rather than absolute perspective. The objective is in fact to identify which areas absorb more costs or require strengthening and thus require action. The analysis can be completed with a comparison (benchmark) with other companies in the same sector or different sectors. The assessment also includes the activities, organisation, technology, competencies, people and the "resources" available.

The evaluation of these variables is linked to various aspects of the model to be implemented. Let us give an example. If the goal is to obtain a flexible model in a certain area, the assessment of the existing resources available for the transformation must be made through this "lens". The main steps of the qualitative evaluation (or self-assessment) of the components could be the following:

1) Definition of a scale for the qualitative assessment of the level of performance of the components, for example from 1 to 5 (insufficient, sufficient, fairly good, good, excellent).
2) Identification of the subdimensions of the flexibility and reactivity analysis of the model: the analysis dimensions are identifiable with the firm's resources, but they can be tailored in relation to the objective.
3) Evaluation and mapping of the results on the CBM map: each component is analysed and evaluated on the dimensions identified and mapping is then performed on the CBM model.

Once the analysis of the identified dimensions has been consolidated, the data are then crossed with the results obtained from the transposition on the map of the flexibility objectives and the short- and medium-term business objectives. The analysis of the results, as mentioned, highlights the areas in which the company must invest to transform its model and the related initiatives. Finally, the components are identified on which to intervene with priority: the so-called "hot" components. The transformation plan is formulated on the basis of the transformation priority of the components (Fig. 4.2).

4.7
Integration in business ecosystems

Firms, in order to operate in the new competitive landscape, must assume a configuration that can be integrated into "open" business ecosystems with an "organic" logic in line with that which Chesbrough, Di Minin and Piccaluga describe in this book (Chap. 3).

Reference can no longer be made to mechanistic models on either the organisational design level or on the management process level. The firm increasingly operates as an organism, in an open environment that also increasingly acts as an organic system. To integrate into this framework, and competitively operate within it, requires changing the models of reference as well as the operating paradigms of the firm. This "organic" analogy entails increasingly referring to the behaviour of natural environments. As concerns the firm, reference should be made to the behaviour of complex animals. As concerns the organisation and its management instead, reference should probably be made to the models of medicine (a discipline that has always had as its objective the interpretation and management of the health of the human body, an organic system).

In fact, in order to identify the most appropriate business model to competitively operate in the expected scenario requires first analyzing from above (with a helicopter overview) the business ecosystem the firm wants to operate in. This would allow identifying the activities and nodes of the ecosystem that can give greater competitive advantage and/or better conditioning in the value chains in which the business is operating. This assessment and analysis of the ecosystem business model

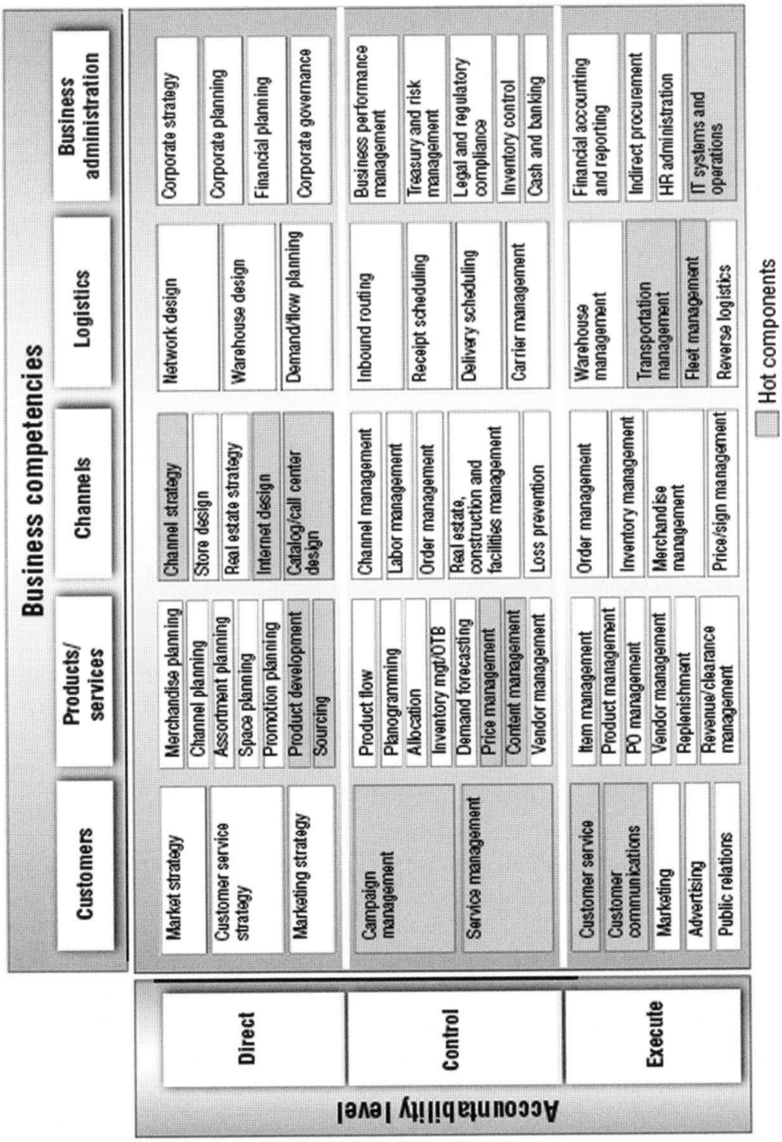

Fig. 4.2 Identifying the "hot components" (http://www-935.ibm.com/services/us/imc/pdf/g510-6163-component-business-models.pdf)

helps in deciding which business roles/components to focus on in order to exploit the scenario in which it operates (also considering the possibilities of creating "entry barriers" for potential competitors).

Already in the 1990s, US Department of Defence and Japanese Ministry of International Trade and Industry studies foresaw what is occurring today in relation to business ecosystems and the functioning of firms within it. These studies assumed traditional value chain management methods based on one to one relationships between suppliers and customers would be surpassed by "almost stable" relations over time. They were to be surpassed by new, more open and dynamic business models. An extensive evolution was foreseen in this sense in the early years of the millennium. And this duly happened, but is now forging ahead with greater speed than anticipated due to technology overboost and the Internet. A key aspect of this development is the transition from the "value chain" logic to that of the "value system" (business ecosystems). This context saw the creation of so-called "business networks" in the global environment. The configurations of advanced networked firms were identified in these studies with new terminology. In Japan the term "holonic" firm was largely used, while in the USA the term "virtual" firm was often used. Other definitions, more or less overlapping, were open business systems, plug-and-run/click and run firms, the extended enterprises and, to some extent, agile production. Today, the term "open business" is more commonly used. In their 1994 book *L'azienda olonico virtuale* [The Holonic Virtual Firm], Merli and Saccani proposed use of the combined name *holonic-virtual system*. The word "holonic" emphasises the importance of a basic structural system to activate new business methodologies. The word "virtual" emphasises the "modus operandi", i.e., how value is operationally created through the combination of participants that operate as one company. The fact that the Japanese mainly used the first term confirms their primary orientation towards prerequisites (the "causes"), while the fact that Americans largely used the term "virtual" confirms their primary orientation towards operability, to operate the "effect". The fact that we Europeans more often use the term "reference model" (which includes both enabling and operational aspects), confirms our Cartesian orientation towards a hierarchical system model approach.

Incidentally, the model implemented by Toyota – when faced with the strategic decision to become world leader in its industry, to simultaneously become a global and multilocal corporation while seeking even greater operational flexibility – contains most of the requirements that are described below.

This is a type of "meta-network" organisation, which involves organisational units and people on a variable basis in function of production volumes required and on the operational contingencies of the business (a decisive change compared to the traditional Japanese closed systems of the previous decade).

The basic assumption of the "holonic-virtual" organic model is that these economic and productive ecosystems require high levels of flexibility and a high degree of autonomy and creativity, even at the operational level, in order to cope with ever-changing markets, customers, environmental, technological and business strategy needs. To meet these requirements, structures are needed that are characterised by the following:

1) articulation of organisations into small operating units;
2) the use of "horizontal" and not hierarchical information networks;
3) the "active" use of the brains of people in all operating positions.

The same group of people must be able to produce and/or integrate the software and hardware of a product/process. The aggregation of people to face problem/opportunity can even happen on long-distance thanks to the Internet (social networks) and intranet, creating ad hoc groups for every need. The same companies can combine in various ways depending on the need of the moment. Thus, a company can be either the customer or supplier of another, a competitor or ally, depending on the business "objective" of the moment and, at times, on the geography of the business. "Small scale" becomes the norm as interaction and mutual dependence of producers and consumers become closer thanks to technology and the Internet. A greater structural capillarity of the system is also required, developing geographic industrial districts that are "virtually" rather than physically proximate. The business organisation is based on a number of interactive "holons" divided into groups and subgroups that can respond creatively to changing market scenarios. The "organic–virtual" company that is created in this context can be defined as follows: "a set of independent operating units acting in an integrated and organic way, as part of an ecosystem, that are configured each time as the most suitable value chain to pursue the business opportunities that the market presents". The "autonomous operating units" mentioned here may be small businesses or parts of companies, but also individuals.

Organic–virtual companies materialise on the basis of market impetuses to meet individual needs or opportunities. They can thus be considered of a permanent type (e.g., related to the life of a product) or a spot type (e.g., to execute an order). They cannot be configured in advance, but only hypothesised, and must develop autonomously based on the contingencies (and over time in "Darwinian" style). This requires taking a pragmatic approach to configuring and developing activities and competencies in the function of the impetuses from the environment and its weak signals. The organic–virtual ecosystem has three basic types of roles:

1) the "key resource" firm;
2) the "operational" firm;
3) the "integrator" firm.

"Key resource" firms have the role of providing the ecosystem with the fundamental factors to activate specific "value chains". These factors, taken individually or combined in different ways, can be identified in the following:

1) developing and managing product and/or service know-how;
2) developing individuals capable of delivering the business objectives;
3) developing and garrisoning/controlling any required specialisations;
4) knowledge of the market;
5) financing.

"Operational" firms are essentially dedicated to the operational execution of the business. These units can be distributed along the entire business chain in "cascade" form (one manufactures and the other sells) or can be vertically integrated in parallel (producing and selling different product lines in parallel). The "operational" firm

Table 4.1 Infomobility ecosystem in CBM form (Mains Laboratory 2008, Pisa)

	The infomobility service system				
	Fleet management: individual transport	Fleet management: freight transport	Traffic and safety management	Mobility payment	Advanced vehicle control
Families of services	Control, monitoring, access services management	Transportation management	Statistical traffic data management	Management, technology and tariff payment integration	Vehicle tracking and telemonitoring
Services	Flexible transport management services	E-supply chain execution	Intelligent traffic light management	Payment	Intelligent navigation system
	Intelligent bus shelter	Field force automation	Limited traffic area access management	Urban and extra-urban road pricing	Integrated vehicle security solutions
	Car sharing and car pooling	Fleet and freight management	Parking management	Rail and local public transport services payment	
			Penalties management	Individual transport services payment	
			Incidental statistical data management	Other city services payment (museums, etc.)	

Table 4.2 CBM analysis of the "Parking Payment" business area (Mains Laboratory 2008, Pisa)

Company	Level	Activities
Mobility management company	DIRECT	Design & planning; Tender; Parking management reporting guidelines; Administrative control; Operational control; Statistical analysis for decisional support; Operational pricing guidelines; Targeted infringement signalling
Auxiliary services company	CONTROL	Penalties monitoring
	EXECUTE	Back office rechargeable cards and subscriptions; Front office for citizens; Emptying parking meters
Sensor system management company	DIRECT	User data management; Statistical data management; Mapping parking management; Starter kit design
	CONTROL	Monitoring; Data processing; Pricing model design; Parking area verification; Additional services; Tariff data dispatch; Position survey
	EXECUTE	Payment activation/ deactivation; Dispatch finance company data; Equipment maintenance/assistance; Sensor system installation; Parking management reporting; Supply data to retail customers; Update variable message panels; Starter kit distribution
IT provider	DIRECT	Software design; Hardware design
	EXECUTE	Software development; Hardware development; System integration; Maintenance assistance IT structure management
Satellite navigation company	EXECUTE	Guide to empty parking space; Booking parking space
Web content provider	EXECUTE	Online map advertising free parking spaces; Booking parking space
Payment management company	EXECUTE	Credit and debt management; Penalties
Bank	DIRECT / EXECUTE	
Others	DIRECT	TELECOMS NETWORK MANAGEMENT guarantee service transport data; GPS INFRA-STRUCTURE MANAGEMENT guarantee service survey/data transport data
	EXECUTE	TELEPHONE CAMPAIGN routing traffic; FIT Distribution "scratch & park"; COURIER Money transportation; ACI et al. Allocation of numbers data interpretation; THIRD PARTIES: geomarketing, etc.

must continuously improve competitiveness in terms of costs, quality, flexibility, reliability and response times in the activities entrusted to it.

"Integrator" firms perform the task of combining the activities of several key resource firms and operational firms (such as integrating the software product manufactured by one company with the hardware produced by another, to then supply it to a commercial unit and thus finalise the value chain). The integrator firm must continually improve its ecosystem knowledge capabilities to ensure maintaining competitive levels and an equitable distribution of margins to the entire chain. Based on this logic, the firm must understand and develop differentiated capabilities to be attractive to the ecosystem.

Bearing in mind these logics and dynamics, a firm must therefore analyze the ecosystems in which they could/would work, identifying potential high-value and highly defendable business roles. The CBM methodology is also well suited to this purpose, first applying it to the business ecosystem and thereafter to the individual firm. Tables 4.1 and 4.2 in this sense show an example of an analysis conducted on the infomobility ecosystem in the Pisa MAINS Master lab. This model can be similarly used to determine how industrial districts can evolve towards a more efficient and competitive logic of business ecosystems operating in the global business sphere.

References

Bauman Z (2002) Modernità liquida. Laterza, Rome
Butera F (2005) Il castello e la rete. Impresa, organizzazioni e professioni nell'europa degli anni '90. Franco Angeli, Milan
Butera F (2009) Il cambiamento organizzativo. Analisi e progettazione. Laterza, Rome
Chesbrough H (2003) Open innovation. Harvard Business Review Press, Cambridge, MA
Galloui F (2002) Innovation in the service economy. Edward Elgar, Cheltenham
Grant R (2010) Contemporary strategy analysis, 7th edn. Wiley, Chichester
IBM Institute for Business Value (2010) Ceo survey 2010: capitalizing on complexity
IBM Institute for Business Value (2008) Ceo survey 2008: the enterprise of the future
IBM Institute for Business Value (2008) Ceo survey 2006: innovation
IBM Institute for Business Value (2010) Cfo survey 2010: the new value integrator
Mchugh P, Merli G, Wheeler B (1996) Beyond business process reengineering. Wiley, Chichester
Merli G (1995) Breakthrough management. Wiley, Chichester
Merli G (1996) Managing by priority. Wiley, Chichester
Merli G (1999) I nuovi paradigmi del management, Il Sole 24 ore, Milan
Merli G (2000) e-biz, come organizzarsi per la net economy, Il Sole 24 ore, Milan
Merli G, Crippa A (2003) Business on demand, Il Sole 24 ore, Milan
Merli G, Saccani C (1994) L'azienda olonico-virtuale, Il Sole 24 ore, Milan
Merli G, Gelosa E, Fregonese M (2010) Surpetere, la competizione creativa efficace e sostenibile. Guerini e Associati, Milan
Rullani E (2010) Modernità sostenibile. Idee, filiere e servizi per uscire dalla crisi. Marsilio, Venice
Valdani E (2009) Client and service management. Egea, Milan

User-led innovation: final users' involvement in value cocreation in services industries

5

Francesco Sandulli

There is widespread consensus among practitioners and scholars alike that end-users can contribute substantially to create new goods and services. Service firms are starting to learn how to involve end-users in the development of new services. Overall, in the case presented in this chapter we see how a company adopted a more open strategy in developing new services. The firm purposely reduced information stickiness by increasing the degrees of interaction between the firm and the end-user and by providing end-users with tools to develop their own applications. The firm also increased the visibility of the implicit benefits of user contributions, reinforcing the organizational identification process by creating a reputational system. These initiatives resulted in better capabilities to detect and filter user demands, but also in greater differentiation from competing investment firms by offering a richer portfolio of services and better analysis and information processing capabilities to end-users.

5.1
User innovation in services firms

During recent years, an increasing number of firms have been changing their innovation models. Traditionally, firms used to adopt closed innovation models, where the knowledge flows that led to product invention, design and production were constrained to the boundaries of the firm. Supporters of a more open innovation process argue that firms need to adopt new innovation models that must involve capable actors not only within the boundaries of the firm, but also outside the firm in order to enhance their innovation capability (Chesbrough 2003). As a result, companies in

F. Sandulli (✉)

Departamento de Organización de Empresa, Universidad Complutense de Madrid, Madrid, Spain

e-mail: sandulli@ccee.ucm.es

L. Cinquini, A. Di Minin, R. Varaldo (eds.), *New Business Models and Value Creation: A Service Science Perspective.* Sxi 8, DOI 10.1007/978-88-470-2838-8_5, © Springer-Verlag Italia 2013

different industries have started to look for other ways to increase the efficiency and effectiveness of their innovation processes through the active search for new technologies and ideas outside of the firm and cooperation with suppliers, competitors and users. The adoption of the open innovation paradigm has been quite heterogeneous. While it is being rapidly adopted by large manufacturing corporations, the current open innovation academic research has not produced enough evidence to convince small manufacturing firms and services firms of any size to abandon their traditional closed model of innovation.

This reluctance in services firms is explained by the fact that much of services innovation is based on experiences, tacit knowledge and intangibles rather than on functionalities, and explicit and structured knowledge, complicating knowledge exchange between services firms. Moreover, services firms are more reluctant to share knowledge because they find more difficulties in achieving replicability, patentability and legal protection for their innovations. However, there are some successful cases of open innovation in services. In this chapter, we will focus our attention on one of these cases in the financial services industry. More precisely, we will study the role of users as cocreators of new financial services.

Starting with the seminal work of Von Hippel (1986), there is widespread consensus among practitioners and scholars alike that end-users can contribute substantially to the creation of goods and services. Firms may obtain some benefits from the participation of end-users in the innovation process. Among others, the literature identifies faster spread of innovations across the users' base, complementing existing innovation portfolios, filling small niches of high need left open by extant firms, reducing information asymmetries between firms and users, gaining access to missing resources (typically knowledge), faster product development and potential cost reductions … (Henkel and Von Hippel 2005; Enkel et al. 2005; Bitzer et al. 2007). Despite these benefits, integrating users in the innovation process can be a challenging task involving transaction and agency costs related to opportunist behaviours of users, causing inefficient product development or excessive orientation towards niche markets … (Athaide and Stump 1999; Brockhoff 2003; Jeppesen 2005; Alam 2006; Gassman et al. 2010). Moreover, in the specific framework of services industries, firms can hardly measure the actual contribution of users to the new service development since service output is not embodied in anything that is physically quantifiable. Measurement biases in relation to services are responsible for the great majority of underestimations of innovation (Gallouj and Savona 2009). If firms cannot properly measure the actual contribution of end-users to the innovation process, they will face high uncertainty when trying to assess the returns on investments of the tools and mechanisms that foster end-user innovation that we are going to discuss in this chapter.

The traditional approach of services firms to user innovation is somewhat limited by the traditional conception of the new service development process as a producer-centred process. In this traditional approach, the main contribution of users to the creation of new services consists of helping firms to discern and deeply understand the users' articulated and unarticulated service-related needs. In this context, firms do not take advantage of users' resources to create and provide new. Consequently, the role of end-users in new services creation has often been limited either to the initial stages of the creation process, idea generation, or to the final stages, testing and validation of the final design of the service. Magnusson (2003) found that user involvement in the generation of new ideas in the telecom industry produced more original ideas, but they were more difficult to implement. Alam (2002) observed that in the financial services industry the most relevant contribution of users to the development of new services will be related to the activities of idea generation and testing of new services, as shown by Thomke (2003) in the case of the Bank of America, which used customers to test new services. Nevertheless, in recent times some services firms are starting to view users as cocreators that can participate actively in the production of new services (Nambisan and Nambisan 2009; Payne et al. 2008; Skiba and Herstatt 2009; Oliveira and Von Hippel 2009; Nambisan and Baron 2010). However, it is not always easy or convenient for firms to engage users in effective cocreation relationships. The degrees of involvement depend on the firm's strategy and the users' willingness to cooperate. The main goal of this chapter is to show with a business case what levers and strategies firms can use to obtain the highest degrees of user involvement in the creation of new services. In the following sections, we will study the influence of information stickiness and expected benefits on end-user willingness to cooperate. In the final two sections, we support our model with a business case in the financial services industry and draw the main conclusions and future lines of research.

5.2
User willingness to cooperate: information stickiness

The innovation literature considers that users' willingness to cooperate is a function of two variables: information stickiness and users' expected benefits (see for instance Von Hippel 1994; Morrison et al. 2000; Franke et al. 2006). Von Hippel (1994) defines the concept of information stickiness of a given unit of information as the incremental expenditure required to transfer that unit of information to a specified locus in a form usable by a given information seeker. Information is sticky because of the way it is encoded or because of the problems of information seekers or providers, such to absorb that information (Cohen and Levinthal 1990). Information stickiness causes users and firms to rely more heavily on information they have in stock than upon information they must draw in from external sources. This in turn means that both users and services firms will tend to develop different *types* of innovations. Users generally have a more accurate and more detailed model of their needs than services firms, while firms have a broader view of the system needed to provide the solution to these needs. This difference is especially clear in those services with a clear distinction between back-office and front-office processes. Users are usually unaware of "what is really going on" in the back-office as they are not in contact with this part of the service system and consequently tend to develop innovations that are functionally novel, more centred on the front-office processes and on the interaction mechanisms between users and firms. These mechanisms tend to require a great deal of user-generated need information and context of use information for their development. In contrast, firms tend to develop innovations that are improvements on well known needs and that require a rich understanding of the whole system for their development (Riggs and von Hippel 1996; Ogawa 1998). When the opacity of the back-office processes to the users is high, the user involvement will be weaker. Von Hippel and Pereira (2009) show this relationship with the case of sweep accounts. Users were only able to develop a prototype or initial solution for some new financial services related to these accounts. The user solution was still far away from the actual solution, since users did not fully understand the adjustments needed to the back-office processes of the banks in order to provide the new service on a large scale. In the process of adapting the user prototype to a real service, the back-office configuration of the banks created some restrictions to the new service design. These limitations produced a nonoptimal solution, which had some differences with the solution proposed by the users since some of the functionality of the user prototypes of sweep services was not included in the final design of the service. Therefore, in this case the banks should have deployed some mechanisms to reduce information stickiness in order to produce a better solution. The literature has mainly identified two of these mechanisms: increasing user–firm interaction or providing users with innovation toolkits.

Regarding user–firm interaction, close interaction between users and firms reduces information stickiness. For instance, studies on Knowledge Intensive Business Services (KIBS) have largely observed this relationship. Within KIBS projects, it is not exceptional that customers do not only explicate their needs but also participate actively in the development or implementation of the solution. Hereby, these services deploy strong interaction mechanisms such as joint teams, face-to-face interactions or training programs to optimize the information flow with users, thereby reducing information stickiness. This interaction explains why user innovation is frequent in KIBS. For instance, Enos (1962) reported that user firms developed nearly all the most important innovations in oil refining. Muller and Zenker (2001) found that manufacturing SMEs interacting with KIBS were more likely to innovate. In turn, the literature also recognizes that KIBS will learn from customers and develop their own innovation capability, leading to new services (Muller and Zenker 2001; Strambach 2001; Fosstenløkken et al. 2003; He and Wong 2009).

Even if effective, intense interaction with users is expensive in terms of internal resources and transaction costs. For this reason, some firms have created another way to reduce information stickiness: toolkits. Toolkits are coordinated sets of "user-friendly" design tools that enable users to develop innovations by themselves. The toolkits are usually specific to the design challenges of a specific field or sub-field, such as integrated circuit design or software product design. With the toolkit, users can create a preliminary design, simulate or prototype it, evaluate its functioning in their own use environment and then iteratively improve it until satisfied. Toolkits reduce the stickiness of a given unit of information by converting some of the firm's expertise from tacit knowledge to a more explicit and easily transferable form (Von Hippel and Katz 2002).

The incentive to invest in reducing the stickiness of a given unit of information will vary according to the number of times that one expects to transfer it; when the number of transfers is high, firms will opt for toolkits. This is the case of telecommunications companies such as Deutsche Telekom, Telefonica, Vodafone or Telecom Italia, among several others that are opening their application programming interfaces (API) and providing their users with software development kits (SDKs) in order to promote the user-led development of applications. In this case, the telecom carrier expects to have the same technical information called on repeatedly to solve n user application problems and that each such problem involves unique user information but similar technical specifications. In this case, the total incentive to unstick the service provider's knowledge across the entire series of user problems is high. Similarly, firms in the financial services industry are starting to embed software in their ATMs that allows customers to create their own customized applications to operate with the bank through ATMs. These two cases show that toolkits are especially useful when service providers attempt to adopt a common and replicable solution approach to diverse application problems of many heterogeneous users. For instance, with the toolkit the telecom company can solve diverse communication-related problems of different users ranging from medical doctors that want to use the open APIs to develop a new mobile e-health application to kids that want to develop a new game for mobile platforms. The commonality in solution approach means that the sticky information required from a service provider to solve each novel application problem tends to be the same, involving such things as the properties and limitations of the solution type. In contrast, the diversity in applications means that sticky information required from users tends to be novel or have novel components. Thus, the higher the heterogeneity of user needs faced by a service provider, the higher its incentive to invest in toolkits (von Hippel 1988). In these contexts of high heterogeneity of user needs, the potential market size is generally small; firms developing toolkits are following a long tail approach to innovation, where they try to address manifold market niches at the same time without incurring high costs, since a significant part of the innovation tasks are performed by users.

5.3
User willingness to cooperate: expected benefits

Users expect to obtain from their innovations a mix of extrinsic/intrinsic collective/individual benefits. These benefits can take the form of money, better service or social recognition. In the specific case of end-users, intrinsic rewards are especially significant (Franke and Shah 2003). Users will be more likely to cooperate in some cases such as when user innovation induces further service improvements by other users or the service providers, the innovation produces a standard solution that is advantageous to the whole base of users, rivalry conditions are low or the user expects positive reciprocity and reputation effects (Harhoff et al. 2000). The Glucoboy case is an example of user-led innovation motivated by intrinsic benefits. This game, developed by Nintendo and launched in Australia on World Diabetes Day in 2009, makes monitoring and achieving blood sugar goals fun. Whenever a user performs a glucose test, points are awarded that allow the user to unlock games. The Glucoboy idea was produced by an end-user, Paul Wessel, a former senior sales executive at Honeywell Corporation, who noticed that his 9-year-old son would constantly deliberately lose his blood glucose meter because he hated testing, but not his Gameboy. Mr. Wessel said that if he could combine blood glucose testing and video gaming technologies, perhaps his son would be more motivated to test. To the end-user, the primary goal of this innovation was not the expected income flows from license sales. In fact, it took three years for Mr. Wessel to convince Nintendo to develop the new service because of the relatively small size of the target market, one of the characteristics of several user innovations. Mr. Wessel said he designed Glucoboy to help other parents avoid what he went through with Luke. He said the new service was designed to make a daily regimen of finger pricking and blood glucose testing, hated by most children who have diabetes, into fun, to motivate them to manage their disease.

In this example, we can see not only the relevance of intrinsic rewards but also the importance of understanding the social context of the end-user. End-users have to be considered by service providers not as isolated individuals, but as individuals in a community of end-users. End-user innovation strongly depends on the value of trust in the relationship with the firm. However, individual trust is strongly influenced by other users' expectations of the relationship with the organization. When the user community is afraid that the firm will act opportunistically and unfairly capture the benefits of the end-user innovation (Bradach and Eccles 1989; Gulati 1995), fostering user innovation will be more difficult. For instance, this is shown by Heiskanen and Lovio (2007) in a case study where engineering firms were unsuccessful in igniting user-led innovation to create new home energy solutions because of the bad reputation of these firms among the user community.

Firms can create trust through an increase in the number of successful interactions with users over time. Through ongoing interaction, companies and end-users learn about each other and develop trust around norms of equity or develop what Shapiro et al. (1992) call "knowledge-based trust". In this sense trust is a way of creating an organisational identification between end-users and the firm. The organisational behaviour literature shows us that social identity plays an important role in the relationship between users and firms. Social identity theory (Tajfel 1978; Tajfel and Turner 1986) posits that individual perceptions and behaviours are guided by social identification processes: the tendency of individuals to think of themselves in terms of the social groups or collectives to which they belong (Tajfel 1978; Tajfel and Turner 1986). Organisational identification denotes self-definition in terms of the membership in the organisation and reflects a sense of unity between self and organisation (Ashforth and Mael 1989; Dutton et al. 1994; van Knippenberg and Sleebos 2006). When the organisation becomes an integrated part of an individual's self-concept, the individual feels included in and part of the organisation.

Organisational identification is important in end-user innovation. Users who identify strongly with an organization are more likely to exert themselves on behalf of the organisation and to show organisational citizenship behaviours, such as proposing innovations or helping the firm to improve its services without being asked to (Ashforth and Mael 1989; Dutton et al. 1994; van Knippenberg and Sleebos 2006).

For this reason, it is important that firms learn how to influence organisational identification. Firms may influence organisational identification by emphasizing the collective in their strategy or displaying collective-oriented behaviour (Shamir et al. 1993). For instance, some firms, such as IBM and SUN, are looking for this identification by supporting the interests of the community of open source developers. Firms may also positively impact feelings of organisational identification by defining a communication strategy aiming at creating the perception among users that procedures and rules that are in use in an organisation are fair and justified (Tyler 1999). Importantly, firms can also strongly influence user identification with the organisation by showing respect for the person and his/her performance (De Cremer and Tyler 2005). In this respect, some firms are starting to create user communities where the main motivation for participation of innovative users is the wish to be recognized by the firm (Jeppesen and Frederiksen 2006). In this case, reputational systems established by the firm are usually positively related to user contributions. Recent research has found that not only do users take on the role of innovators, but they may also establish and organize the innovation communities themselves (Lettl and Gemünden 2005).

5.4
Business case: reducing information stickiness and reinforcing organisational identification in the financial services industry

In today's volatile economy, successful financial services professionals are quickly responding to a fluctuating market by creating agile and flexible dashboards to analyse market performance, trading portfolios and risk analysis. In this section we will analyse how a firm in the financial services industry leveraged on the power of toolkits to reduce information stickiness and community benefits to foster the capability of end-users to create new services.

InvestGreat Group is a European financial services company founded in 1987 that manages more than €6 billion through six mutual funds, 33 SICAVs and three pension funds, and has more than 32,000 customers, including several financial advisors. The firm achieved a strong brand and market recognition due to solid performance of their funds and SICAVs during the stock market crash of 2000–2002. In 2002 their Spanish equities fund obtained an 8% return vs. a 28% fall by the Spanish stock market. The firm currently has the second largest Spanish equity fund. Despite this track record of good results, the latest financial turmoil seriously affected the firm. It lost nearly 13,000 customers in less than two years and its funds lost between 45% and 3% of their value, the highest losses in the history of the firm, positioning it among the ten worst investment funds in Spain. The firm's assets value was reduced from nearly €4 billion to €1.5 billion in 15 months.

After the worst part of the crisis was over, the InvestGreat CEO decided to focus on the optimisation of the internal operations of the firm, neglected during the fast growth years prior to the crisis. In the "good times" the management of the firm was strongly focused on the maximisation of the financial value of the assets managed. Asset managers of the firm spent 95% of their time searching for companies and assessing their objective value. To that end, they used an extensive number of sources: visits to companies, industry publications, contacts with other similar managers abroad, foreign press, analyst reports, etc. To determine the objective value they took into consideration both key business features (cyclical, stable, capital-intensive, etc.) and the skills and honesty of their management teams. However, during the crisis they discovered that this strong focus on asset valuation had meant that the firm performed some commercial and administrative processes badly, especially customer relationship management, considered secondary to the core financial analysis process. Among some other initiatives, the firm focused on innovation and on reinforcing the relationship with customers. In this respect, two projects were started in 2009: an online portal to facilitate customer operations and the creation of an on-line community of users to improve customer interactions.

Through the community, which is accessible only by invitation, the firm tries to establish a closer relationship with its clients. The online community allowed users to take advantage of both the knowledge, which other investors have decided to share, and the outside information, which the firm made available to them. Investors in the community will receive in their personal home pages opinions from other members of the community, stock prices provided by Thomson Reuters and financial papers, and other news from economic papers. The community also provided users with tools to develop customised services to manage their investment information. In fact, via secure connections to the websites of their banks, the firm employs a financial aggregator that, using bank-level security, pulls together all the user investment accounts and puts each operation into flexible categories. But the killer feature is showing where the average return of the user's investments are below the odds relative to other similar users. Besides, users can easily and rapidly deploy by themselves new customised applications to manage information regarding their investments. Some examples of the new services developed by some users are the historical valuation of their investments, daily updates of the value of their investments, regular e-mail or SMS alerts on the status of their investments, applications to organise investment information into different portfolios irrespective of the banks where they are held and customised analytical tools to compare the performance of their investments with leading market indices, among other new services. Not only the customer that has created the service will benefit from its innovation, but the rest of the community will also. In fact, the firm has started to scrutinise the applications developed by users, looking for new services that would be of interest to replicate and make accessible to the rest of the community.

The distinctive point regarding this case is the fact that InvestGreat combined toolkits and community management. Some other firms provided just toolkits, such as the Automated Trading of XTB and the investor community XtraInvestor. The development of the community of users falls into the strategies we have discussed above to increase end-user innovation. Firstly, the community reduces information stickiness, reducing the opacity of the back-office processes of the firm. In this respect, InvestGreat managers now share with their customers the information sources used to manage their funds. On the other hand, users can use their miniblogs to make their needs more explicit. Moreover, InvestGreat managers now have the means to filter these needs. In fact, they can identify the most relevant new service demands by filtering by the number of users asking for it, the number of users who have developed their own application to solve it or the type of user demanding it, since the firm has developed a reputation algorithm to detect the lead users. Secondly, the community has reduced information stickiness by providing users with toolkits to develop their own applications and services. As explained in the theoretical background, the firm has applied a common solution approach, the aggregation of information from banks, to a heterogeneous set of customers with different demands. During the first six months of operations of the community, more than 300 customers developed their own applications. The firm will subsequently study which of these applications might be interesting to scale up to the whole base of customers. The community manager identifies those applications that might be of interest for the rest of the community and, together with the IT staff, the technical suitability of the new application. Thirdly, the community reinforced the trust of their customers in the firm. Information stickiness has been reduced, since investors know what information sources are being used to make investment decisions with their money. While previously communication between GreatInvest managers was reduced to a monthly newsletter or the content on the website, in the new user community managers have their own blogs, where they and users can share opinions almost on a daily base. This new communication channel increases the degree of interaction between users and the firm and consequently the levels of customer trust. Finally, we have observed that the community also allowed the organisational identification of end-users with the firm. Users can have their own blogs with a GreatInvest appearance and feel, creating a visual identification between users and the firm. In the first six months, around 1200 users created blogs and started sharing their opinions or information with other users. Moreover, to promote user contributions, the firm included a reputation algorithm based on the comments users posted on other users' personal blogs. The empirical analysis of contributions has shown that users with higher reputations do contribute more to the community and that, in turn, increases in reputation scores for a given user increase the likelihood that this user will make significant contributions to the community. As a global result of these trust-building initiatives, a first preliminary result of the implementation of the new strategy was the reduction by two points of the churn rate compared to the average in the previous four semesters. From the point of view of new services, the firm is now offering two new services developed by users: an application to develop a graph of the internal rate of return and a global investment position graph summarising portfolio valuation with the option of considering or not market valuation.

5.5
Conclusions

Firms can benefit from the contributions of end-users to create new products. These contributions are more significant for the firm's business model when users are highly motivated and involved in the development process. To optimise user involvement, firms have to deploy strategies to reduce information stickiness and to increase the benefits expected to be obtained from the innovation. Firms can reduce information stickiness with intense interactions with users. However, these interactions are costly. For this reason, firms are starting to embed knowledge in toolkits, which are especially efficient in those cases where firms can employ constrained and homogeneous technical information to solve a set of heterogeneous needs from different users. Past studies have shown that the participation of users in the creation of new products highly depends on the users' expected intrinsic benefits and trust on the firm. Moreover, firms must understand that users are not isolated individuals but members of a community. Strategies building trust and favouring intrinsic benefits must be deployed at the community level rather than at the individual level. Overall, in the case we have described above, the firm adopted a more open strategy in developing new services. The firm purposely reduced information stickiness by increasing the degrees of interaction between the firm and the end-user and by providing end-users with tools to develop their own applications. The firm also reinforced the visibility of the implicit benefits of user contributions and the organisational identification process by creating a reputational system. These initiatives resulted in better capabilities to detect and filter user demands, but also in greater differentiation from competing investment firms by offering a richer portfolio of services and stronger analysis and information processing capabilities to end-users. Despite the strong synergies derived from combining toolkit strategies and community-building strategies, firms must be aware of the special initial conditions that need to apply in order to maximise the returns of this combination. First, firms adopting this combined strategy may solve heterogeneous needs for a set of users that share some characteristics and therefore join up a community. That is, users must share an organisational identity. If users are heterogeneous, it will be more difficult to create a common organisational identity, limiting the replicability and scalability of new products or services and the capitalisation of intrinsic benefits. Second, firms should be able to solve the heterogeneous needs of users using a homogeneous set of knowledge and resources. If firms need to apply different knowledge or resources to the different needs, toolkits and a community-based approach are not efficient, it being more advisable to follow one-to-one interactions with users. Third, firms must be conscious that user innovation usually leads to new products that are of interest only for niche markets. For this reason, firms must adopt a long tail approach to innovation strategy, at least for open innovation processes. This long tail approach can be appropriate only if firms and users share risk, profits and resources to create new products. Users are a barely utilised source of knowledge and resources that firms must start learning to exploit in a new competitive framework of continuing reinvention of their business model.

References

Alam I (2002) An exploratory investigation of user involvement in new service development. J Acad Mark Sci 30(3):250–261

Alam I (2006) Removing the fuzziness from the fuzzy front-end of service innovations through customer interactions. Ind Mark Manag 35(4):468–480

Ashforth BE, Mael F (1989) Social identity theory and the organization. Acad Manag Rev 14(1):20–39

Athaide GA, Stump RL (1999) A taxonomy of relationship approaches during product development in technology-based, industrial markets. J Prod Innov Manag 16(5):469–482

Bitzer J, Schrettl W, Schröder PJH (2007) Intrinsic motivation in open source software development. J Comp Econ 35:160–169

Bradach JL, Eccles RG (1989) Price, authority, and trust: from ideal types to plural forms. Annu Rev Sociol 15:97–118

Brockhoff K (2003) Customers' perspectives of involvement in new product development. Int J Technol Manag 26(5–6):464–481

Chesbrough H (2003) Open innovation: the new imperative for creating and profiting from technology. Harvard Business School Press, Cambridge

Cohen W, Levinthal D (1990) Absorptive capacity: a new perspective on learning and innovation. Admin Sci Q 35(1):128–152

De Cremer D, Tyler TR (2005) Managing group behaviour: the interplay between fairness, self, and cooperation. Adv Exp Soc Psychol 37:151–218

de Jong JPJ, Vermeulen PAM (2003) Organising successful new service development: a literature review. Manag Decis 41(9):844–858

Dutton JE, Dukerich JM, Harquail CV (1994) Organizational images and member identification. Admin Sci Q 39(2):239–263

Enkel E, Kausch C, Gassmann O (2005) Managing the risk of customer integration. Eur Manag J 23(2):203–213

Enos JL (1962) Petroleum progress and profits: a history of process innovation. MIT Press, Cambridge, MA

Fosstenløkken SM, Løwendahl BR, Revang Ø (2003) Knowledge development through client interaction: a comparative study. Organ Stud 24(6):859–879

Franke N, Shah S (2003) How communities support innovative activities: an exploration of assistance and sharing among end-users. Res Policy 32(1):157–178

Franke N, von Hippel E, Schreier M (2006) Finding commercially attractive user innovations: a test of lead user theory. J Prod Innov Manag 23(4):301–315

Gallouj F, Savona M (2009) Innovation in services: a review of the debate and a research agenda. J Evol Econ 19(2):149–172

Gassman O, Kausch C, Enkel E (2010) Negative side effect of customer integration. Int J Technol Manag 50(1):43–62

Gulati R (1995) Does familiarity breed trust: the implications of repeated ties for contractual choice in alliances. Acad Manag J 38(1):85–112

Harhoff D, Henkel J, von Hippel E (2000) Profiting from voluntary information spillovers: how users benefit from freely revealing their innovations. MIT Sloan School of Management WP #4125, July 25

He ZL, Wong PK (2009) Knowledge interaction with manufacturing clients and innovation of knowledge-intensive business service firms. Innov Manag Policy Pract 11(3):264–278

Heiskanen E, Lovio R (2007) User knowledge in housing energy innovations. Proceedings of the Nordic Consumer Policy Conference, Helsinki, October 3–5

Henkel J, von Hippel E (2005) Welfare implications of user innovation. J Technol Transf 30(2_2):73–87

Jeppesen LB (2005) User toolkits for innovation: consumers support each other. J Prod Innov Manag 22(4):347–362

Jeppesen LB, Frederiksen L (2006) Why do users contribute to firm-hosted user communities? The case of computer-controlled music instruments. Organ Sci 17:45–63

Lettl C, Gemünden HG (2005) The entrepreneurial role of innovative users. J Bus Ind Mark 20:339–346

Magnusson PR (2003) Managing user involvement in service innovation: experiments with innovating end-users. J Serv Res 6(2):111–124

Morrison PD, Roberts JH, von Hippel E (2000) Determinants of user innovation and innovation sharing in a local market. Manag Sci 46(12):1513–1527

Muller E, Zenker A (2001) Business services as actors of knowledge transformation: the role of KIBS in regional and national innovation systems. Res Policy 30(9):1501–1516

Nambisan S, Baron R (2010) Different roles, different strokes: organizing virtual customer environments to promote two types of customer contributions. Organ Sci 21(2):554–572

Nambisan P, Nambisan S (2009) Models of consumer value co-creation in healthcare. Health Care Manag Rev 34(4):334–343

Ogawa S (1998) Does sticky information affect the locus of innovation? Evidence from the Japanese convenience-store industry. Res Policy 26(7/8):777–790

Oliveira P, von Hippel EA (2009) Users as service innovators: the case of banking services. Research Paper, MIT Sloan School of Management

Payne AF, Storbacka K, Frow P (2008) Managing the co-creation of value. J Acad Mark Sci 36(1):83–96

Riggs W, von Hippel E (1996) A lead user study of electronic home banking services: lessons from the learning curve. MIT Sloan School of Management Working Paper

Shamir B, House R, Arthur MB (1993) The motivational effects of charismatic leadership: a self-concept based theory. Organ Sci 4(4):577–594

Shapiro D, Sheppard BH, Cheraskin L (1992) Business on a handshake. Negot J 8(4):365–377

Skiba F, Herstatt C (2009) Users as sources for radical service innovations: opportunities from collaboration with service lead users. Int J Serv Technol Manag 12(3):317–337

Strambach S (2001) Innovation processes and the role of knowledge-intensive business services (KIBS). In: Koschatzky K, Kulicke M, Zenker A (eds) Innovation networks. Concepts and challenges in the European perspective. Physica-Verlag, Heidelberg

Tajfel H (ed.) (1978) Differentiation between social groups: studies in the social psychology of intergroup relations. Academic Press, London

Tajfel H, Turner JC (1986) The social identity theory of inter-group behavior. In: Worchel S, Austin LW (eds), Psychology of intergroup relations. Nelson-Hall, Chicago

Thomke S (2003) R & D comes to services: Bank of America's pathbreaking experiments. Harvard Bus Rev 81(4):71–79

Tyler TR (1999) Why people cooperate with organizations: an identity-based perspective. Res Organ Behav 21:201–246

van Knippenberg D, Sleebos E (2006) Organizational identification versus organizational commitment: self-definition, social exchange, and job attitudes. J Organ Behav 27(5):571–584

von Hippel E (1986) Lead users: a source of novel product concepts. Manag Sci 32(7):791–805

von Hippel E (1988) The sources of innovation. Oxford University Press, London
von Hippel E (1994) Sticky information and the locus of problem solving: implications for innovation. Manag Sci 40(4):429–439
von Hippel E, Katz R (2002) Shifting innovation to users via toolkits. Manag Sci 48(7):821–833

INNOVATION LAB

Marketing and strategy: the economic impact of the development of an NGN on the country system. Telco and broadcast tv convergence scenarios

MAINS Master, academic year 2008/2009
People and companies involved in the InnoLab:
Students: Giulia Crespi, Fabrizio Falchi and Giacomo Sorbi
Companies: Ericsson Telecomunicazioni, Telecom Italia and Vodafone
Professors: Daniele Dalli and Riccardo Lanzara

1. The problem

New generation network (NGN) commonly refers to the creation and development of telecommunication networks towards a common typology that allows transportation and delivery of all TLC services (voice, data and multimedia communications, games, advertising, etc.), encapsulating information packets and, in most cases, entrusting their management to the IP protocol.

More specifically, these packet communication networks (such as the Internet, differing from the old analogue circuit-switched telephone networks) are able to provide services regardless of the technologies underlying the broadband infrastructure. An NGN can offer access not only to individual users, but also to different service providers (such as advertisers, broadcasters and public administrations); the service supply should extend from the landline (the old fixed-line networks) to the mobile (network connectivity developed today mainly for mobile phones).

From an architectural point of view, the implementation of an NGN entails significant costs: coverage of about 50% of the Italian population in the fibre to the home (FTTH) mode is currently estimated at a cost of €13 billion over 10 years. A lively debate is also underway about the operating modes to be selected depending on the maintenance costs, the quality of the service or the actual possibility of unbundling the offers from the divergent solutions that have involved incumbent Telecom Italia and other major carriers such as Fastweb, Tiscali, Vodafone and Wind.

Subsequently, smaller players (known as "OLOs" or other licensed operators) have also come forward, proposing a consortium of all operators and the possibility of having shareholdings in direct proportion to the use of the network, subject of course to the possibility of buying, renting or selling shares between operators.

Ministerial and private sources estimate the impact of such an operation in the range of between 40 and 420 billion in increased net sales for the entire supply chain in over 10 years, as well as the creation of 250,000–300,000 new IT jobs.

In light of both the costs and the possible consequences, and given the practical impossibility of creating multiple "parallel" NGNs, the critical issue of opportunely defining standards and common-use strategies as well as developing products and business formulas to make the most of this opportunity is evident, which is bound to increasingly affect GDP, also considering that Italy is among the last to progress compared to the rest of the EU, which already has cabling, dissemination and commercial experience of a certain level.

2. Work methodology

Following the preliminary phase of market, technology, operator and underlying infrastructure recognition, the team focused on a more specific study of the development and marketing of formulae and products related to internet protocol television (IPTV), or rather, cable TV that leverages not on an ad hoc network but on that used for internet data transmission.

Among the prime criticalities that directly emerged were:

1) the need to find market experiences as closely related to the Italian situation as possible in order to be able to study its evolution;
2) the role and importance of the various operators in the industry both on an organizational/creative level and on the extended value-chain level;
3) the possibility of implementing multiple business models, not necessarily in mutual exclusion, but almost certainly more or less profitable for one or other link in the value chain.

The first part of the work focused on the analysis of other successful experiences in the introduction and diffusion of IPTV, making a selection of the most interesting cases.

Working towards macro areas, the team identified as particularly interesting and paradigmatic some players in the United States, the world's largest market, with a strong tradition in traditional cable that is slowly but surely being converted to NGN. The analysis of the French market, at the time the largest IPTV market in the world, was indispensable, also due to the pioneering gambles of a few of the operators analysed.

The working group then analysed the behaviour of big IPTV players in Japan and South Korea, distant countries in many respects, but similar in some econometric features and the morphology of the territory on which to implement NGN. This first phase of work was further subjected to filters and refinements.

All information examined was thus collected and analysed in light of the contingencies and the Italian market where, for example, a single plan for the transition to NGN had not yet been firmly delineated and in which perhaps the closest experience, satellite pay-per-view, took around 5 years to reach the quota of 1 million subscribers.

From this preceding stage, new insights and new questions emerged, such as:

1) Who could the key Italian player be to steer and organise the entire process towards the development and better use of NGN?
2) What has the outcome and feedback of players involved in the modest IPTV trials undertaken to date revealed?
3) Which technologies and what timescale are involved in adapting the network to transmission standards and the provision of ancillary services?
4) What kind of devices?
5) Marketed under which formulae?
6) Who and how much to charge?
7) How to divide revenue throughout the value chain?
8) Which newcomers, with respect to the TLC industry, could enter the market?

In order to respond to these questions, the team followed three main guidelines: First, the group benefited from the support of CNIT professionals and researchers who provided a more complete and updated picture of the opportunities and challenges in the Italian scenario.

Second, interviews were conducted with other professionals, such as a lecturer and author of NGN, operators of the network infrastructure, competitors of business partners and national broadcasters. Another particularly relevant experience was the conversation with the head of new technologies of a major national television network and the discovery that formats and devices had already been studied for a number of years to enhance the interaction with end-user opportunities. The issue of piracy is particularly thorny, especially considering the qualitative level of the transmission of television and cinema products, which is equivalent or even superior to that obtained from buying or renting the same products on optical media.

Also emerging from the interviews of opinion leaders were several concerns about the operational, legal and economic issues of the IPTV offer: value distribution, business mode, additional services, greater or lesser transparency with respect to devices, piracy and so forth. These are only some of the pitfalls to overcome to achieve a goal that is still on the distant horizon, offering great potential even with the current small user base, but also evidencing the first problems to be surmounted.

The third path of the team's activities, alongside interviews with big names in the IPTV arena, consisted in the creation of a panel of around ten lead users who were given a standard questionnaire to identify the salient characteristics: starting from factors such as awareness of the reality of respondents and their circle of peers, proposing different scenarios for their use, with different technological and pricing approaches and services provided.

All data collected were further examined in a final meeting between the students of the Master, tutors and academic partners to develop the definitive elaboration of the research results.

3. Proposed solution

The end result of the research was a provisional, and thus strategic, marketing proposal with related revenue forecasts within a certain range of diffusion over the next 5 years.

Following the forecast of attainable subscribers, explicit assumptions were also formulated on the commercial formulae and the relative industry turnover to be allocated according to the business models to be created and proposed to the market.

Finally, a less "numerical" but more strategic analysis was undertaken of the many criticalities that emerged in the course of the work, especially in interviews with opinion leaders and lead users. The solutions to some of these (consider, for example, advertising management, the creation of specific formats for IPTV capabilities or the implementation of effective measures to regulate the protection of copyrights) probably went beyond the original purpose, time and resources of the lab, but are likely to be the subject of future research studies as a consequence of their criticality.

Models of performance and value measurement in service systems

6

Roberto Barontini, Lino Cinquini, Riccardo Giannetti, and Andrea Tenucci

The emergence of service science offers new and renewed research interest to management accounting and to performance management. In terms of management accounting, the prospects of cocreation of value and servitisation lead towards analysis objectives that consider the customer to a greater extent. Secondly, the trend of dissociation between investments (costs) and sources of revenues questions the validity of the traditional logic of costing for pricing in the context of service science. In short, problems of cost and revenue allocation emerge between coproducing partners in a service system. In terms of performance management, the development of business models in which the relevance of the service component increases requires reflection on which innovative techniques should be used to measure value, also with a view to establishing incentive systems oriented towards value creation.

6.1
"Service" as the object of measurement

As widely illustrated in Chap. 1, the service concept has various dimensions and interpretations, and has been subjected to an evolution on both macroeconomic and

R. Barontini
Istituto di Management, Scuola Superiore Sant'Anna, Pisa, Italy
e-mail: r.barontini@sssup.it

L. Cinquini (✉)
Istituto di Management, Scuola Superiore Sant'Anna, Pisa, Italy
e-mail: l.cinquini@sssup.it

R. Giannetti
Dipartimento di Economia e Management, Università di Pisa, Pisa, Italy
e-mail: rgiannet@ec.unipi.it

A. Tenucci
Istituto di Management, Scuola Superiore Sant'Anna, Pisa, Italy
e-mail: a.tenucci@sssup.it

L. Cinquini, A. Di Minin, R. Varaldo (eds.), *New Business Models and Value Creation: A Service Science Perspective.* Sxi 8, DOI 10.1007/978-88-470-2838-8_6, © Springer-Verlag Italia 2013

microeconomic levels. The evolution of this concept in recent years is closely related to the extensive and pervasive impact of the diffusion of Information and Communication Technology (ICT) in all fields of human activity. In the sphere of economic activity, ICT has revolutionised the concepts of space and time, of "product" and of the organisational methods to obtain it (Normann 1996). Emerging technologies have opened the new world of economies of "increasing returns" of products based on knowledge, profoundly different from the traditional "decreasing returns" of tangible products, where the strategic, operational and management rules are very different (Arthur 1996).

To frame the problem from the analysis of this chapter, we classify the interpretations of "service" in the existing literature into four categories (Normann 1984; Grönroos 1998, 2008), which are briefly presented in a sequence that expresses the evolution of the concept over time:

1) service defined on the basis of discriminating characteristics in relation to goods;
2) service as a process or activity;
3) service as a business perspective;
4) the system of services from the service science perspective.

6.1.1
The "four main characteristics" (IHIP) as differentiating factors between goods and services

Research on services is historically based on features distinguishing between services and physical goods. Generally, a distinction that constitutes a key reference point is ascribable to Shostack (1977), who described the differences between products (goods) and services from the marketing management perspective in four aspects: intangibility, heterogeneity, inseparability and perishability – from which the acronym IHIP derives.

1) Intangibility: Services are intangible and insubstantive, cannot be touched, seized or manipulated. In this sense, they do not require transport, storage or stockpiling.
2) Heterogeneity: Each service is unique. It is instantly generated, rendered and consumed, and can never be exactly repeated since the point in time, place, circumstances, conditions and current configurations of the resources allocated are different every time, even if the same customer requests the same service. ICT facilitates the standardisation of certain services (e.g., Internet-based services and telecommunications or ATM banking) and this characteristic can therefore not be generalised (Lovelock and Gummesson 2004).
3) Inseparability: Service providers are essential to service delivery; they must promptly generate and render the service to those consumers requesting it. In many cases, service provision is performed automatically, but the service provider must first allocate resources and systems and actively maintain adequate response capacity and efficiency in the service provision. In addition, service consumption is inseparable from its supply because the customer is involved

in its delivery, from the initial request up to service consumption. In addition, service production and its consumption by the consumer occur simultaneously.[1]

4) Perishability: Services are perishable in two respects:

a) The resources, processes and systems relevant for the provision of a service are assigned in a given period of time. If the customer does not request consumption of the service in this period, it cannot be redirected to other customers. From the perspective of the service provider, this determines the criticism of the planning phase of the use of available capacity, since the service cannot be produced in advance and stored to be sold later. For example, an empty seat on a plane can never be used after takeoff, a hotel cannot recover lost revenue from the non-use of rooms and a consultant without customers forever loses the possibility to bill the time that has elapsed.

b) When the service has been completely rendered to the consumer who requests the service, it disappears irreversibly, since it has been consumed by the customer. For example, a passenger has been transported to his destination and other customers cannot be transported again to that destination at that same time.

Over the past decade, the IHIP characteristics have often been criticised due to their non-systematic development as well as their complexity and subjectivity (Grönroos 1998; Lovelock and Gummesson 2004; Vargo and Lusch 2004). The fundamental problem is that IHIP characteristics are presented as a service definition when in reality they are only a set of characteristics, often loosely coupled (Laine et al. 2009).

6.1.2
Service as a process or activity

Service is also interpreted as a process or series of actions (activities) that distinguish it from physical products (output), namely, the overall actions taken to provide a solution to a customer's problem (Grönroos 2008). According to Hill (1977), service means "a change in the condition or state of an economic entity (or object) caused by another person or entity. [...] A service activity is an operation designed to bring about a change in the state of a reality C that is owned by [a customer] B and implemented by a service provider A" (Araujo and Spring 2006). In this definition, service appears to be rather focused in relation to processes and customer needs; the service provider implements these processes to enable the desired changes in the customer's world. To be noted is that the reality C, owned by the customer, can consist not only in goods but also in "conditions" or "status" that can be improved by the service provider.

---

[1] On this point, however, Lovelock and Gummesson (2004) noted that many services are partially or fully produced independently from the customer, who is not directly involved (consider, for example, car repairs, information, financial services, transport of goods), with the result that production and consumption may not be simultaneous.

It is worth noting, however, that customers consume services, irrespective of whether buying "goods" or "services" (Vargo and Lusch 2004, 2008). According to this approach, customers can create value only through the consumption of services, which in the case of goods is implicit in their use.

6.1.3
Service as a business logic: service-dominant logic

The recent service-dominant approach broadens the interpretation of service. Vargo and Lusch (2004), referring in particular to marketing studies, sustain that "New perspectives converge to form a new dominant logic for marketing, where the provision of services rather than goods is fundamental for economic exchange". Relationships with customers, intangible goods and cocreating value with the customer, in fact, form the basis of the production of services (Vargo and Lusch 2004). In this perspective, all firms are ultimately service companies, thus obeying the *service-dominant logic*. The quality of customer interaction is fundamental to the competitive success of the contemporary firm. The concept of cocreating value is intended as customer participation and consideration of their contribution in one of the stages that characterise the implementation of the offer (value proposition): design, implementation and post-sales. Involvement can have consequences for the exploration of trends, the creation of a need and its satisfaction, and contributes to the acceptance and success of a firm's value proposition (Merli et al. 2010, pp. 128 ff.).

Vargo and Lusch (2004, 2008) define services as "the application of specialised competencies (knowledge and ability) through actions, processes and performance to benefit another entity or the entity itself". In this approach, the service (rather than the product) creates a benefit for the customer and goods are interpreted accordingly as mere tools or mechanisms for the distribution of the service provision. Service, therefore, constitutes the general case, the common denominator of the exchange process: in reality, it is service that is always exchanged, while goods, when they are used, constitute a support in the service delivery process.

Consequently, Vargo and Lusch (2008) make some fundamental assertions as the basis of the emergent logic that is dominated by service (service-dominant logic), which is set against the traditional logic that is dominated by goods (goods-dominant logic):

- In terms of the economic system, service is the fundamental basis of exchange (services are exchanged for services, service represents affluence, a physical good is only its wrapping); physical goods are only a mechanisms for the distribution of services, in the sense that they derive their value from use, or rather, from the service they provide. Hence all economies are fundamentally service economies.
- Customers are cocreators of value, in terms of the interactive process with the service provider. The latter are unable to create value by themselves, but collaboratively offer their resources and contribute to value following their acceptance of thiscooperation . In other words, the firm can only make a "value proposi-

tion" to customers and their agreement is the condition for the effective creation
of value.
- Service is determined by the interaction with the customer and, therefore, inherently oriented to the fulfilment of specific needs.

6.1.4
The service system from the Service Science perspective

The concept of service, as shown in recent literature, cannot be defined clearly and
unambiguously, perhaps as a result of heterogeneity between the different characteristics of different types of services.

However, several elements that are common to different types of services have
been identified as (Chesbrough and Spohrer 2006):

- the general existence of close interactions between supplier and customer;
- the nature of knowledge created and exchanged;
- the simultaneity of production and consumption;
- the combination of knowledge that is determined within systems following the
  interaction of their components;
- exchange as a process and as an experience;
- the role of ICT and its contribution to the transparency of the exchange process.

The speed and extent of change in the economic and technological environment,
in which services undoubtedly have an increasingly relevant role, have forced both
rethinking the approaches to business management issues and defining the requirements in terms of competencies needed for their governance.[2] In this context, in 2003
IBM brought a new approach to the fore. Pivoting on service as a central object of
modern business configurations, an integration of science, engineering and management was proposed to enable effective solutions to modern business problems: the
key objects of analysis in the new context become "service systems" (IBM 2004).

A service system is any entity that produces and consumes services, considering
both the internal structure and the external ecosystem of services in which they exist
and interact (Spohrer 2008; Katzan 2008). In this perspective, four types of entities
can be observed: households, businesses, cities and nations. In turn, in each of these,
a considerable number of internal structures interact; they are referred to as "components" or "internal service systems"[3]. A service system is thus a configuration of
coproduction of value consisting of (1) people, (2) technologies, (3) other internal

---

[2] We can summarise the three basic trends characterising the evolution of firms' strategic and management
scenarios (Merli 2003): (a) changes in the structure of the business relating to structural relationships between firms (network); (b) changes in business scenarios that relate to globalisation, convergence (of new
actors working on the platform of the global network with new ways of collaborating), technology and the
role of finance; and (c) the evolution of e-businesses related to business opportunities generated/fuelled
by new technological opportunities. See also Freeman (2005).
[3] For example "Business exists in a complex ecosystem of service exchange. Businesses have a considerable amount of internal structures, which allows a business to be viewed as a set of components or internal
service systems." (Spohrer 2008: 31). See Chap. 1 and Chap. 4.

or external service systems and (4) shared information (language or norms) (Maglio and Spohrer 2008). Design, implementation and management in terms of continuous improvement of services require specialised competencies in organisational change (human factors), business design (economic and management factors) and the design and implementation of technology (engineering factors).

Among the conditions that render the realisation of service systems possible, two in particular are of fundamental importance: modularity and standardisation (Baldwin and Clark 2000; Gallinaro 2009). These concepts, borrowed from the industrial arena, can be applied to products/services or organisations: in the latter perspective "a modular organization is a modular system of loosely coupled activities and processes ? these same processes being partly independent and partly interdependent" (Gallinaro 2009, p. 7). In particular, the modularity of processes is applied to the inter-firm dimension, achieved through the use of common platforms, or links modules belonging to different organisations, through standardised interfaces (Pekkarinen and Ulkuniemi 2008).[4]

Considering the abovementioned elements, from the following questions some fundamental research issues arise in the service science perspective, with implications for performance measurement (IfM and IBM 2008):

1) How can the different architectures and basic components explain the origins, the life cycle and the sustainability of the performance of service systems?
2) How can the functioning of service systems be optimised in order to interact and cocreate value?
3) How and under what conditions do certain interactions that intervene within and between service systems lead to specific results?

Given the complexity of services as "the object of measure", the investigation that we conduct in the next few sections focuses on processes, systems and business models that are particularly innovative, as well as on their operation and management logic.

6.2
Management accounting and the measurement of service performance

The issue of service in its most modern definition progressively emerged in management accounting research starting in the early 1990s. Traditionally management studies focused on manufacturing activity, since this environment in fact constitutes the original development context of cost and management control tools.

With respect to the applicability of the traditional approaches developed in management accounting for decision support and management control in services, the main limitations linked to the characteristics mentioned in the preceding section concern (Amigoni 1986, 2000; Dearden 1978; Modell 1996; Fitzgerald et al. 1998):

---

[4] A transposition in terms of the strategic analysis of the modularity of service systems is made through the *Component Business Modelling* (CBM) approach (see Chap. 4).

1) The uselessness of cost allocation on the final product aimed at valorising inventories due to the absence of stocks in service companies; therefore, no classic distinction is made between "product costs" and "period costs".

2) The difficulty in measuring output in qualitative and quantitative terms as a result of its intangibility, which is significantly complicated with respect to both physical assets and customer evaluation of benefits, as well as the effective value of resources used for the services provided. Their measurement necessarily involves reference to the activities and processes implemented in the service provision and often their specificity with respect to the customer makes them widely disparate.

3) The intangibility of the fundamental asset used in services, human capital, and the difficulty of expressing it in monetary terms, with significant consequences on measuring the economic value of service companies.

4) The "process of cocreation of value" and the involvement of customers within the process introduces strong elements of uncertainty in the design and management of control systems due to the behaviour of customers and the definition of the boundaries of internal responsibility. The variability of needs and expectations of customers induce variability in the response of the service firm; the ambiguity in the actual controllability of the processes results in significant difficulties in assessing individual performance. These aspects reflect on the efficiency of planning and control systems. Furthermore, the interaction between different organisations in implementing service systems poses the problem of control on an interorganisational level.

The evolution in cost and performance measurement tools (consider for example activity-based costing and the Balanced Scorecard) has been, and is, an opportunity for service companies to innovate and improve management accounting. These approaches internalise on one hand the focus on business activities and processeswhich is – fundamental for the analysis and governance of services, – and on the other, they integrate non-financial metrics with economic metrics for correct identification of the ultimate causes (drivers) of performance. This supports the organisation's orientation towards the objectives and for an effective evaluation of the results (Brimson and Antos 1994; Kaplan and Cooper 1998; Kaplan and Norton 2008).

The evolution of service systems in the perspective outlined above calls for management accounting research, particularly with respect to some important areas that we describe below.

First, the cocreation of value and servitisation perspective (with respect to manufacture, see Chap. 2), emphasises the role of the customer. Not only is the "cost of production" (process) or "product cost" (output) alone relevant but also the "cost of use" within the overall life cycle of the product. In other words, the focus shifts to the analysis of the costs of the services offered by the artefact (physical product) and their maintenance over time. It is thus possible to sustain innovative design strategies of the offer that link the costs incurred (or sustained) to the user's utility (Normann 2001, pp. 150–153).

In this perspective, costing systems capable of identifying the "total cost of ownership" (TCO) are increasingly relevant. TCO analysis was created with reference to a more accurate assessment of costs in the supply chain, enabling understanding of the burden beyond the transaction price for the duration of the use of the good or service.[5] However, this approach can also be applied with respect to the final customer, to understand the nature and efficiency of services provided by the manufacturer/supplier and to act in terms of both performance improvements and in terms of the efficiency of the cocreation of value. This evolution is linked to the progressive shift of the strategic focus (especially for industrial companies) from the physical production processes to the processes of use of what is produced (Normann 1984). The shift in the cost valorisation objective (from the product to its use) can support decisions aimed at reducing the cost of use for the customer through innovating the design of the offer. This way, the performance/cost ratio can improve and hence the value for the customer; this benefit can be quantified by determining the reduction of the monetary outflow of the customer who owns and consumes the product/service. In the ElsagDatamat case described below, TCO is presented as an instrument that facilitates interaction between customers and suppliers for the formation of the commercial relationship as well as improving performance characteristics over time.

Case study

ElsagDatamat is the centre of excellence of the Finmeccanica Group for specific technologies and competencies in ICT, automation and security. It also offers integrated control and physical security solutions for civilian markets (oil and gas, transportation, metro and railways, power generation, marine and offshore) as well as logical security solutions. ElsagDatamat adopted TCO in 1999–2000, driven by the need for an instrument that would allow them to obtain information to evaluate the advantages of outsourcing decisions.

Currently TCO is used to manage relationships with suppliers and customers. The former entails an internal use of TCO (typical) to acquire a broad understanding of supply costs beyond the purchase cost of the various production factors. The latter concerns customer relationships. ElsagDatamat's services, for example, SAP Hosting, offer the customer the alternative of continuing to perform these services on their own or delegate them to ElsagDatamat.

---

[5] TCO is a cost management tool that aims at determining for the firm, as acquirer, those costs considered relevant or significant in the acquisition, possession, use and elimination of a good or service. With respect to a supplier, the buyer's TCO includes not only the price of the goods/services purchased but also the costs for order management, research and supplier qualification, transportation, receipt, inspection and any eventual restitutions, storage and disposal. This analysis can be relevant not only to assess supply relationships but also in terms of outsourcing decisions. On the characteristics and application methods of this instrument, see Ellram (1993), Ellram and Siferd (1998) and Pitzalis (2009).

TCO in this case allows communicating to the customer the economic benefits arising from the choice of attributing the implementation of certain services to ElsagDatamat. TCO means customers can be shown the costs they would sustain if continuing to operate the service on their own in comparison to outsourcing the service to ElsagDatamat.

ElsagDatamat, therefore, offers their customers a comparison, over a period of three or five years, between the TCO, or the cost, that they would continue to sustain if they chose to keep "possession" of the service and the cost if they delegated it to ElsagDatamat. TCO strives to take account of items and cost configurations (such as energy and environmental costs) that the customer would most likely neglect if performing a traditional and differential "make or buy" type analysis.

Determining the TCO "in the customer's perspective" can be provided by ElsagDatamat with different degrees of involvement: ranging from situations where the customer is almost entirely willing to share information on costs with ElsagDatamat in order to be able to determine the TCO, to situations where, instead, ElsagDatamat estimates costs based on knowledge of the business sector. TCO, in both the uses described above, provides useful information for potential cost reductions. More specifically, by expressing the weight that the specific cost items/aggregates have on the TCO, in addition to understanding how their incidence changes over time, allows them to obtain useful information to guide initiatives to reduce TCO (obviously acting on the variables controlled by the firm such as the choice of suppliers).

Among the developments that ElsagDatamat foresees for TCO, of particular interest is the adaption of this instrument to cloud computing. Cloud computing refers to the customer's remote use of hardware and software using, for example, software that resides on a server without the need to download it. This results in interesting TCO application perspectives on both the customer (service user) and supplier side since both parties are interested in knowing how the cost of owning certain ICT services changes with this type of configuration.

A second aspect is the increasingly frequent dissociation of investments (costs) and sources of revenues that come to light with respect to services provided to customers by the most innovative network platforms, posing new problems in the traditional costing for pricing logic (Bhimani and Bromwich 2010). In such contexts, pricing and costing do not follow traditional cost-plus or market-based type models. Pricing is instead linked to the business strategy and revenue generation dynamics, which are increasingly dissociated from the product cost that is being measured. In Internet-based services (e.g., social networks or marketplace platforms such as eBay), digital products include functionality but also personal entertainment that the consumer experiences in the use of the service. Often these platforms are instrumental in the independent creation of a "product" by the customer in the cocreation logic (e.g., Facebook).

The volume of users in these cases becomes the most important driver of revenues and profitability in terms attraction of advertising investments, targeted at grabbing the customer's attention. Thus, the conditions are created to cover the volume of investments (fixed costs) in ICT infrastructures related to the development of software applications. However, the direct link between costs and prices, which characterises the business logic of firms embedded in the goods dominant logic, is lost. In traditional manufacturing activities, the causal cost-pricing link derives from focusing on the direct processes that create products or services, under the assumption that the producers should manage the resources they own. In the world of services, instead, it is the process of value cocreation that is the ultimate source of profitability and must be monitored and managed.

In particular, if we consider Internet-based service firms, we can observe that the control systems which predict future revenues (and not so much costs) are preeminent, since the value of these firms is mainly reflected by the number of customers and the related stream of future revenues (Sjöblom 2003). This perspective comprises product or service price discounting phenomena, at times innovative, that may ultimately be offered free of charge (Amigoni 2000): the attraction of a substantial market share is fundamental in the network services business (e.g., a browser or a search engine such as Google) and the price is sometimes a simple tool to attract consumer attention to an undifferentiated product ("commodified") (Bertini and Wathieu 2010). Both cases go far beyond the traditional logic of *costing for pricing*.

A third key area in which we believe service science influences the development of management accounting research is the problem of the distribution and measurement of costs and revenues between coproducers/cocreators of value in a system of services. In this context, the aspect of cocreating value with customers once again becomes relevant, in terms of both business-to-business and business-to-consumer relationships. Value no longer derives from internal firm efficiency alone and is calculable, in Porter's (1980) value chain perspective, as the difference between sales revenues and the cost of "strategically relevant" activities constituting the chain. In the service-oriented logic, the firm's objective is *mutual value creation* for the firm as well as its customers while service is the mediating factor in this process (Grönroos and Ravald 2009). In other words, the value that a firm can create from the customer relationship depends on the value that customers themselves can create from involvement in the relation. We refer to mutual value creation in this sense: the customer acts as coproducer within the supplier's process and the supplier simultaneously acts within the corresponding process of creating customer value and is actively involved therein (Grönroos and Helle 2010, p. 570). Grönroos and Helle (2010) make also a step forward towards measuring value in this logic by a model where the evaluation considers joint supplier–customer productivity and how this derives from the efficiency and effectiveness of the relationship itself. It is clear that this measurement depends on the availability of data based on costs and cash flows as well as the degree of trust and the willingness of the actors involved to share their accounts.[6]

With a view similar to the "mutual value creation" logic, Pardo et al. (2006) argue that there are three categories of value: *exchange value*, which originates in the firm's activities and is consumed by the customer; *proprietary value*, exclusively created and consumed by the firm since it carries out activities according to efficiency and effectiveness criteria; and *relational value*, cocreated by the customer and the firm as a result of activities that straddle the boundaries of the two actors. It is this latter case that affects relationship performance, a measure of value created in the customer–supplier relationship over time, determining how this performance is shared between the firm (in the form of value capture) and the customer (value creation). With this in mind, and focusing on value captured by the service provider, Storbacka and Nenonen (2009) suggest that value capture can be measured as the actualisation of future profits arising from the relationship with the customer. They also argue that this value can be used as a proxy for shareholder value creation. The value of a long-term customer–supplier relationship (Ravald and Grönroos 1996), especially in service firms, becomes not only a marketing objective, but also an issue to which management accounting can substantially contribute.[7]

6.3
The role of technology and data analysis processes in management accounting of services

Finally, it seems relevant to consider some possible impacts that ICT, along with data analysis processes, could have on management accounting systems and the measurement/management of performance used in services. As noted, the progression and diffusion of ICT has enabled encoding, archiving, processing and making available for subsequent applications a quantity of knowledge that was previously tacit since it is incorporated in the behaviour of people and/or artefacts (Chesbrough and Spohrer 2006). The increase of available knowledge allows the design and development of new services; furthermore, computer technology, together with quantitative and qualitative methods of analysis, allows the performance of data and information analyses that can significantly contribute to the generation of competitive advantages and their sustainability. *Analytics* (or analytical information) is the extensive use of data and explanatory and predictive analyses through statistical and econometric models to make decisions and perform actions (Davenport and Harris 2007, p. 7).[8] Naturally, analytics concerns not only services "in the strict sense", but also goods

---

[6] In this sense, the terms interorganisational cost management (Cooper and Slagmulder 1999; Hoffjan and Kruse 2006) and open book accounting (Hakansson and Lind 2007; Giannetti 2009a) are often referred to.

[7] The role of management accounting in service companies should therefore go beyond the simple measurement of internal efficiency that characterises production companies (Lowry 1993).

[8] Analytics is part of business intelligence technologies and processes that handle data to understand and analyse the performance of organisations (Smith and Goddard 2008, p. 128).

that, as previously mentioned, can be considered a means to enable service provision. That said, we will consider some effects that analytics can have on management accounting systems and service performance measurement with reference to the development of new services, the design and use of performance measurement systems and cost management tools.

Analytics can be usefully employed in the development of new services, opening up interesting integration scenarios with management accounting studies. As noted, service innovation is a particularly critical phase, since, amongst other things, it determines the essential characteristics of the service and thus the value creation potential.[9] Therefore, the adoption of measurement and performance management systems useful for assessing the "productivity" of this process is considerably interesting. However, studies on the use of performance measures to determine the outcome of the development of new services has been relatively scarce (Storey and Kelly 2001), despite the fact that services, with respect to material goods, have distinctive features (Tatikonda and Zeitham 2002). Davenport and Harris (2007, p. 76) point out that analytics can be useful at this stage to identify possible areas of innovation and simulate the impact of proposed changes on future activities. This information can obviously be very useful in assessing the viability of new services and also, more generally, in improving the performance of the service development process. Moreover, it is likely that the process management may require the development of monetary and non-monetary measures whose integration could benefit, as will be seen later, from the contributions that analytics can provide.

More generally, with regard to performance measurement systems, research confirms the progressive use of non-monetary indicators alongside traditional monetary indicators (Holloway 2009). The literature (Scapens and Bromwich 2010), as also mentioned in preceding paragraphs, has widely demonstrated the importance assumed by multi-dimensional performance measurement systems (Amigoni and Miolo Vitali 2003). The performance measurement of services confirms this trend, although with a different relevance among the adopted measures in companies, according to the contingent variables affecting the specific business context (Brignall and Ballantine 1996).

Non-monetary and monetary measures can be integrated in various ways (Pitzalis 2003), but some research suggests that it is not easy to find a structured link between such measures in performance measurement systems adopted in practice (Ittner and Larcker 2003). One can easily imagine how such a link could instead be useful to verify, in order to shed light on the contribution of non-economic and financial drivers to economic value creation.[10] Analytics can also be used for this purpose (Davenport and Harris 2007, pp. 61–62) and can furthermore provide useful information to identify the determinants of "hidden" value and to assist in price fixing and the management of production constraints, decisions that typically involve both management accounting and performance management.[11]

---

[9] See, in particular, Verweire and Revollo (2009).

[10] For a critical analysis of these links, see also Nørreklit and Mitchell (2007).

[11] Davenport and Harris (2007, p. 43), for example, demonstrate the case of a hotel company that, through analytics, defined prices to optimise the exploitation of available production capacity.

With regard to cost management, given the process-nature of services, it seems important to highlight the possible link between so-called activity-based tools (activity-based costing/management) (Kaplan and Cooper 1998) and analytics. The critical role of ICT in the effectiveness of these instruments has already been highlighted (Kaplan and Anderson 2007); here we want to stress the benefits that the use of analytics can offer to cost management. The "mathematics" of cost allocation is not in itself complex, while in managing services can be more difficult (Davenport and Harris 2007, p. 64) to: (1) identify, disclose and use the non-monetary data needed to allocate costs (allocation bases) and (2) implement analyses/simulations[12] using cost information obtained to select the most significant cost drivers and verify the impact on the profitability of alternative decisions (such as changes to the offer to certain customer segments selected through an analysis of their level of loyalty). Analytics can contribute significantly in these phases.[13]

In sum, analytics can produce useful information to refine – in a way that is unique and difficult to repeat – the efficiency and effectiveness of development processes and service provision, thus contributing to the generation of competitive advantage.[14] The application of analytics to services seems to delineate interesting prospects of integration with management accounting, performance measurement and management systems. Of course, these prospects need to be confirmed by appropriate empirical evidence that will help to understand the contexts in which analytics can be used with greater success.[15]

6.4
Value measurement and incentive design in the service systems

As mentioned previously, a particular characteristic of high value-added services is strong customer involvement in value creation. Innovating by providing advanced services to businesses and consumers entails the development of a relationship based

---

[12] See, as an example, the analysis on cost drivers carried out by Banker and Johnston (1993) to examine the impact on costs of the volume and complexity/variety of services offered.

[13] The use of appropriate costing systems can be useful to "discover" the value created by certain specific groups of customers who buy specific bundles of services. It should be noted that "components" or modules of services can be a source of value if they are proposed as a unique offer to specific customers. For the analysis of the profitability of customers, time-driven activity-based costing can also be used: through time equations, it seems particularly suited to simulate the economic effects resulting from the personalisation of certain types of services. On customers as a fundamental object of analysis of the value created in services, see also Collini (2006) and Cugini et al. (2007). On time-driven activity-based costing, see Kaplan and Anderson (2007). It is worth mentioning that this variant of activity-based costing, given its relatively recent introduction, has not yet seen a great deal of useful empirical evidence to assess its potential effects. See also Giannetti (2009b).

[14] A process must clearly be part of a valid business model to allow the benefits of a sophisticated analysis of the information to be reaped.

[15] Davenport and Harris (2007, p. 36), for example, note that in a sample of 32 companies, most of those that show a high intensity of analytics adoption designed to generate and maintain competitive advantage are companies with a high intensity of information content, belonging, however, to different sectors.

on the exchange of knowledge and an active customer role in finding solutions to their specific needs. The close relationship between supplier and service user is the basis for long-term collaboration, which usually extends well beyond the single transaction.

This section will developed the implications determined by these characteristics on (a) the definition of objectives in terms of value creation (b) the methodologies able to measure results, and (c) incentive systems most suitable to induce managers to invest more efficiently in the service sector.

The first aspect to consider concerns the choice of objective function that managers should seek to maximise. Does the classic criterion of maximising firm value in the long term continue to be appropriate in a context of value cocreation? In fact, it could be argued that the greater role of customers should lead managers to assume decisions taking into account the interests of other stakeholders.[16]

With regard to this aspect, Jensen (2010) examines the apparent contradiction between the objective of maximising shareholder value and stakeholder theory. Strictly speaking, the idea of taking into account the value created by the customer is not incompatible with the objective of maximising firm value. It is clear that, in order to create shareholder value, managers must obtain the active cooperation of their partners within the system of services offered; however, if the firm operates in a sufficiently competitive system and if there are no relevant externalities,[17] the capacity of the business model to generate cash flows (even if not directly from customers[18]) retains its validity not only in relation to the service provider but also as a signal of the overall perception of the value obtained. In other words, maximising the value jointly created can be translated for both the supplier and the user of the service into specific objectives and in individual assessments of the value created.

In advanced business to business services the specificity could rather be determined by the importance of controlling the conflicts between the objectives of the two "partner" firms (supplier and user),: excessive uncertainty surrounding the management of the relationship could slow down the decision-making process and penalise the competitive capacity of the firms involved. One way to align the incentives of suppliers and users is to offer payment for the service linked to performance. Hypko et al. (2010) propose a review of common practices used by manufacturing firms to manage contractual relationships taking into account the performance ob-

---

[16] See, for example Brignall and Ballantine (1996) and Lapierre (1997).

[17] Of course, this hypothesis could not hold, in particular within the service sector.. If we consider, for example, the rating sector, i.e., the evaluation of the creditworthiness of companies, we see that the market is dominated by a few operators and has high entry barriers. In this case, high prices can maximise the value of rating agencies, at the detriment of their customers; similarly, the reduction of control activities by rating agencies may determine costs to the community that may be difficult to overturn on the same credit rating agencies (Hill 2004). In general, in the service sector, it would seem particularly important to increase competition, for the benefit of users and weaker partners (see, for example, the debate on the "Bolkestein" European Directive No 2006/123/EC).

[18] Consider, for example, all web-based services that do not involve any payment for the user but whose volume of traffic creates the conditions for obtaining revenues.

tained by the customer; this work demonstrates the diffusion of pricing schemes based on results achieved by the service user, such as cost savings achieved, the increase in revenue generated or the change in profit margin (Hünerberg and Hüttmann 2003; Helander and Möller 2008).

Turning to the methodologies able to measure the results obtained, it is worthwhile to observe firstly that the application of the criterion of value maximisation in the services sector does not imply the uselessness of a system of indicators that helps managers to set specific goals. Even if at top management level the use of parameters relating to market return as an indicator of value creation within each organisation is acceptable, this general criterion must be concretely declined into the operational objectives that each employee can understand and manage. In this framework, maximising value and the use of systems such as the Balanced Scorecard (Kaplan and Norton 2008) are not only compatible, but can be synergic: the process of defining the best drivers to use, i.e. those more related to the purpose of maximising value in the long term, gives the opportunity to reflect on the most important strategic levers able to achieve this goal.

However, some specificities of advanced services complicate the identification of drivers that can guide decisions and promptly monitor the dynamics of the firm, before the effects of decisions become irreversibly evident in the realisation of cash flows. As mentioned earlier, services are often:

1) heterogeneous, since they are linked to the specific characteristics of the user;
2) related to intangible and not easily valuable assets;
3) marked by significant integration between the firms involved;
4) characterised by high-risk and a medium–long time horizon.

One of the challenges facing service science literature is therefore not only to identify how innovation in services can provide a competitive advantage to the firm, but also in the suggestion on how service value could be effectively transferred in new metrics.

As previously mentioned, the approach proposed by Helle (2009, 2010) and Grönroos and Helle (2010) identifies value cocreation in the interaction between firms, focusing on drivers related to joint cash flow generation.[19] This entails, however, significant measurement problems, not only in terms of the considerable detail of accounting data on costs incurred by the two firms involved, but especially for the generation of future cash flows determined by the service.[20] An innovative service can in fact lead to business opportunities that are difficult to estimate and that

---

[19] In particular, this is first defined in terms of incremental cash flow that the service user derives in terms of new revenues and costs, to which is added the cost incurred by the service provider. The modality of sharing the value created between the partners by defining the pricing of the service is only considered secondarily.

[20] The example used by Grönroos and Helle (2010, pp. 581–584) shows a very simplified case of outsourcing services: the evaluation of the new service is implemented mainly in terms of cost savings while the expected increase in revenues is considered as a certain value. The difficulty of assessing the value created for the customer can often lead to inadequate pricing of the service (Rapaccini and Visintin 2009).

are furthermore characterised by a significant risk profile (Neely 2008), which is not easily treatable in the sphere of traditional evaluation methods.[21]

The correct evaluation of a new service requires not only adequate information, but also new metrics and a new accounting logic. To date, the issue does not seem to have been addressed with particular attention by the literature[22] and does not find immediate acceptance in accounting practices, with the effect of making the contribution of services to value creation scarcely visible. This could lead also to investment disincentives (Kerr 2008): if the costs incurred for the development of these activities reduce the operating profit, since they are not recognised as new assets, managers may be induced to invest sub-optimally for fear of being penalised by the reactions of the market or the firm's lenders.[23]

We must recognise the considerable difficulty in identifying, within the sphere of a strict accounting logic, the value of intangible assets, since this information on future business opportunities is inherently subjective and not immediately verifiable. Its quantification may lead to further conflicts between management and shareholders,[24] which can only be partially limited by the search for robust evaluation methods.

If, therefore, current accounting practices face the difficulty of credibly disclose the value linked to business models identified by service science, then a particularly important role should be recognised for the use of incentive systems, as these are potentially able to induce management to make decisions aimed at creating value (Jensen and Murphy 1990).

However, particular attention must be paid to the design of incentive systems for managers, as has been clearly highlighted by the recent financial crisis: among the causes that led to this situation are the excessive and distorted incentive systems based on stock options. These instruments, which should have aligned the incentives of bank managers with those of shareholders, led to excessive risk-taking and sometimes gave rise to actual fraud.[25]

---

[21] The estimate of expected cash flows does not take into account the flexibility of firm behaviour, which is certainly relevant in the context of innovative services, that have a substantial technological dimension. Evaluation methods should consider the value of flexibility, which could be estimated with the methodology of real options.

[22] For example, the survey proposed by Sharma and Kumar (2010) on the use of EVA does not identify, in the vast literature analysed, studies that specifically address the issue of performance measurement in the field of advanced services.

[23] In this perspective, it would be appropriate for accounting practices to comply with this agency problem, in response to economic changes also linked to the development of service science (Mattessich 2006). For example, methods similar to those used in the insurance industry to estimate embedded value, which links firm value to the expected value of the margins of the previously acquired policy portfolio, coud be used for advanced services. According to this technique, the value of the portfolio in place is determined by estimating the present value of the expected subsequent profits that the portfolio will generate over its remaining life.

[24] In concentrated firm ownership, the conflict would occur between shareholder-managers and firm creditors.

[25] See, in particular, Hall and Murphy (2003), Kedia and Philippon (2009) and Johnson et al. (2009).

As Bebchuk and Fried (2004) suggested, incentive systems should be defined coherently with the objective to encourage managers to maximise firm value in the long run. In particular, they should:

1) identify the specific contribution of managers to the improvement of firm performance;[26]
2) consider a long-term time horizon;[27]
3) provide not only incentives but also penalties related to the failure of achieving a minimum target.

In response to the financial crisis, for example, the banking system has clearly defined guidelines for the definition of incentive systems actually focused on value creation.[28]

A challenge awaiting the services sector in the next few years is the voluntary establishment – beyond impulses deriving from the regulation – of methods identifying value and a way of defining incentive systems that capture the specificity of the advanced service systems, in order to induce actions aimed at creating value. Some empiric results[29] reveal that performance measurement systems are capable, even in the service sector, of better aligning operational activities with the strategic objectives of the firm and their use is positively correlated to the performance achieved.

References

Amigoni F (1986) Il controllo di gestione nelle imprese di servizi. Sviluppo e Organizzazione 95(May–June):7–16

Amigoni F (2000) I costi, dalla produzione di massa all'economia dell'informazione. Economia & Management, July

Amigoni F, Miolo Vitali P (eds) (2003) Misure multiple di performance. Egea, Milan

Araujo L, Spring M (2006) Services, products, and the institutional structure of production. Ind Mark Manag 35:797–805

Arthur WB (1996) Increasing returns and the new world of business. Harvard Bus Rev July–Aug:100–109

---

[26] This suggests the use of performance parameters calculated as the difference with respect to the general trend, or, preferably, the sector of belonging. For top managers , market-oriented performance parameters should be adjusted by the effect of exogenous factors, outside of their control, while for operational roles bonuses should reflect the specific contribution to improving the performance of the firm.

[27] A short-term horizon of incentive schemes can induce decisions at odds with the perspective of maximising long term value and also incentivise accounting manipulations aimed at increasing short-term performance.

[28] In line with the criteria of the Financial Stability Board (2009), the Bank of Italy (2009) explained the guidelines for the definition of incentives in Italian banks. In particular, it requires that bonus payments be deferred over time and linked to long-term performance indicators, appropriately adjusted for risks assumed. It also suggests that remuneration be symmetric with respect to the results achieved, therefore considering penalties for underperformance and takes into consideration the results at business unit and, where possible, individual level.

[29] See, for example, Evans (2004).

Baldwin CY, Clark KB (2000) Design rules: the power of modularity. MIT Press, Cambridge, MA

Bank of Italy (2009) Sistemi di remunerazione e incentivazione, 28 ottobre

Banker RD, Johnston HH (1993) An empirical study of cost drivers in the U.S. airline industry. Accounting Rev 68(3):576–601

Bebchuk L, Fried J (2004) Pay without performance: the un-fulfilled promise of executive compensation. Harvard University Press, Cambridge

Bertini M, Wathieu L (2010) How to stop customers from fixating on price. Harvard Bus Rev May:84–91

Bhimani A, Bromwich M (2010) Management accounting: retrospect and prospect. CIMA Publishing

Brignall S, Ballantine J (1996) Performance measurement in service businesses revisited. Int J Serv Ind Manag 7(1):6–31

Brimson JA, Antos J (1994) Activity-based management for service industries, government entities and nonprofit organizations. Wiley, New York

Chesbrough H, Spohrer J (2006) A research manifesto for services science. Commun ACM 49(7):35–40

Collini P (2006) Cost analysis in the hotel industry: an ABC customer focused approach and the case of joint revenues. In: Harris P, Mongiello M (eds) Accounting and financial management. Elsevier, Oxford

Cooper R, Slagmulder R (1999) Strategic costing and special studies. Strat Finance 80(5): 14–15

Cugini A, Carù A, Zerbini F (2007) The cost of customer satisfaction: a framework for strategic cost management in service industries. Eur Accounting Rev 16:499–530

Davenport TH, Harris JC (2007) Competing on analytics. The new science of winning. Harvard Business School Press, Boston

Dearden J (1978) Cost accounting comes to service industries. Harvard Business Review September–October:132–140

Ellram LM (1993) Total cost of ownership: elements and implementation. Int J Purch Mater Manag 29(4):3–10

Ellram LM, Siferd SP (1998) Total cost of ownership: a key concept in strategic cost management decisions. J Bus Logist 19(1):55–84

Evans JR (2004) An exploratory study of performance measurement systems and relationships with performance results. J Oper Manag 22(3):219–232

Financial Stability Board (2009) FSB principles for sound compensation practices. Implementation Standards, 25 September 2009

Fitzgerald L, Johnston R, Brignall S et al (1998) Misurare la performance nelle imprese di servizi. Egea, Milan

Freeman TL (2005) The world is flat. Farrar, Straus and Giroux, New York

Gallinaro S (2009) La modularità nello sviluppo e nella produzione dei servizi. Impresa Progetto 1:1–22

Giannetti R (2009a) L'Open Book Accounting. In: Miolo Vitali P (ed.) Strumenti per l'analisi dei costi. Percorsi di Cost Management, vol III. Giappichelli, Turin

Giannetti R (2009b) Il Time-driven Activity-based Costing. In Miolo Vitali P (ed.) Strumenti per l'analisi dei costi. Approfondimenti di Cost Accounting, vol III. Giappichelli, Turin

Grönroos C (1998) Management e marketing dei servizi. ISEDI, Turin

Grönroos C (2008) Service logic revisited: who creates value? And who co-creates? Eur Bus Rev 20(4):298–314

Grönroos C, Helle P (2010) Adopting a service logic in manufacturing. Conceptual foundation and metrics for mutual value creation. J Serv Manag 21(5):564–590

Grönroos C, Ravald A (2009) Marketing and the logic of service: value facilitation, value creation and co-creation and their marketing implications. Working Paper 542, Hanken School of Economics, Helsinki

Hakansson H, Lind J (2007) Accounting in an interorganizational setting. In: Chapman CS, Hopwood AG, Shields MD (eds) Handbook of management accounting research, vol 2. Elsevier, Oxford

Hall BJ, Murphy K (2003) The trouble with stock options. J Econ Perspect 17(3):49–70

Helander A, Möller K (2008) System supplier's roles from equipment supplier to performance provider. J Bus Ind Mark 23(8):577–585

Helle P (2009) Towards understanding value creation from the point of view of service provision. Working paper, Conference Report EIASM Service Marketing Forum, Capri

Helle P (2010) Re-conceptualizing value-creation: from industrial business logic to service business logic. Working paper. Hanken School of Economics

Hill CA (2004) Regulating the rating agencies. Washington Univ Law Q 82:43–95

Hill TP (1977) On goods and services. Rev Income Wealth 23(4):315–338

Hoffjan A, Kruse H (2006) Open book accounting in supply chains: when and how it is used in practice? Cost Manag November–December:40–46

Holloway J (2009) Performance management from multiple perspectives: taking stock. Int J Prod Perf Manag 58(4):391–399

Hünerberg R, Hüttmann A (2003) Performance as a basis for price-setting in the capital goods industry: concepts and empirical evidence. Eur Manag J 21(6):717–730

Hypko P, Tilebein M, Gleich R (2010) Clarifying the concept of performance-based contracting in manufacturing industries. A research synthesis. J Serv Manag 21 (5):625–655

IBM (2004) IBM Research. Service science. A new academic discipline? Paper in http://www.almaden.ibm.com/asr/SSME/

IfM and IBM (2008) Succeeding through service innovation: a service perspective for education, research, business and government. University of Cambridge Institute for Manufacturing, Cambridge, UK

Ittner CD, Larcker DF (2003) Coming up short on nonfinancial performance measurement. Harvard Bus Rev November:88–95

Jensen MC (2010) Value maximization, stakeholder theory, and the corporate objective function. J Appl Corp Finance 22(1):32–43

Jensen MC, Murphy KJ (1990) Performance pay and top-management incentives. J Pol Econ 98(2):225–264

Johnson S, Ryan HE, Tian YS (2009) Managerial compensation and corporate fraud: the sources of incentives matter. Rev Finan 13:115–145

Kaplan RS, Anderson SR (2007) Time-driven activity-based costing. Harvard Business School Press, Boston

Kaplan RS, Cooper R (1998) Cost and effect: using integrated cost systems to drive profitability and performance. Harvard Business School Press, Boston

Kaplan RS, Norton DP (2008) Execution premium. Harvard Business School Press, Boston

Katzan H (2008) Service science. iUniverse, New York

Kedia S, Philippon T (2009) The economics of fraudulent accounting. Rev Finan Stud 22(6):2169–2199

Kerr SG (2008) Service science and accounting. J Serv Sci 1(2):17–26

Laine T, Paranko J, Suomala P (2009) All activities are interpretive: the end of the debate about service characteristics? Paper presented at the 2009 Naples Forum on Service: Service-Dominant Logic, Service Science, and Network Theory, Capri, 16–19 June

Lapierre J (1997) What does value mean in business-to-business professional services? Int J Serv Ind Manag 8(5):377–397

Lovelock C, Gummesson E (2004) Whither services marketing? In search of a paradigm and fresh perspectives. J Serv Res 7(1):20–41

Lowry J (1993) Management accounting's diminishing post-industrial relevance: Johnson and Kaplan revisited. Accounting Bus Res 23(90):169–170

Maglio PP, Spohrer J (2008) Fundamentals of service science. J Acad Mark Sci 36(1):18–20

Mattessich R (2006) The information economic perspective of accounting: its coming of age. Can Accounting Perspect 5(2):209–226

Merli G (2003) Business on demand. Il prossimo paradigma. Come vincere nel nuovo scenario competitivo. Il Sole 24 Ore, Milan

Merli G, Gelosa E, Fregonese M (2010) Surpetere, la competizione creativa efficace e sostenibile. Guerini e Associati, Milan

Modell S (1996) Management accounting and control in services: structural and behavioural perspectives. Int J Serv Ind Manag 7(2):57–80

Neely A (2008) Exploring the financial consequences of the servitization of manufacturing, Oper Manag Res 1(2):103–118

Normann R (1984) Service management: strategy and leadership in service businesses. Wiley, Chichester

Normann R (1996) Services in the neo-industrial society. 8th Convegno di Sinergie: "L'impresa e il management dei servizi nell'economia neo-industriale", Naples, 18 October

Normann R (2001) Reframing business: when the map changes the landscape. Wiley, Chichester

Nørreklit H, Mitchell F (2007) The balanced scorecard. In: Hopper, T, Northcott D, Scapens R (eds) Issues in management accounting, 3rd edn. Prentice Hall, London

Pardo C, Henneberg SC, Mouzas S, Naudè P (2006) Unpicking the meaning of value in key account management. Eur J Mark 40(11/12):1360–1374

Pekkarinen S, Ulkuniemi P (2008) Modularity in developing business services by platform approach. Int J Logist Manag 19(1):84–103

Pitzalis A (2003) L'integrazione delle informazioni: una review sulle ricerche empiriche. In: Amigoni F, Miolo Vitali P (eds) Misure Multiple di performance. Egea, Milan

Pitzalis A (2009) Il Total Cost of Ownership. In: Miolo Vitali P (ed.) Strumenti per l'analisi dei costi. Percorsi di Cost Management, vol III. Giappichelli, Turin

Porter M (1980) Competitive strategy. Free Press, New York

Rapaccini M, Visintin F (2009) In search of a product-service strategy. ASAP Service Management Forum, 5–6 November, Brescia

Ravald, A, Grönroos, C (1996) The value concept in marketing. European Journal of Marketing, 30 (2): 19–30.

Scapens RW, Bromwich M (2010) Management accounting research: twenty years on. Manag Accounting Res 21(4):278–284

Sharma AK, Kumar S (2010) Economic value added (EVA): literature review and relevant issues. Int J Econ Finance 2(2):200–220

Shostack GL (1977) Breaking free from product marketing. J Mark Theory Pract 41(2):73–80

Sjöblom L (2003) Management accounting in the new economy: the rationale for irrational controls. In Bhimani A (ed.) Management accounting in the digital economy. Oxford University Press, Oxford

Smith PC, Goddard M (2008) Performance management and operational research: a marriage made in heaven? In: Thorpe, R, Holloway J (eds) Performance management: multidisciplinary perspectives. Palgrave MacMillan, New York

Spohrer J (2008) Service sciences, management and engineering (SSME) and its relation to academic disciplines. In: Stauss B, Engelmann K, Kremer A, Luhn A (eds) Service science. Fundamentals, challenges and future developments. Springer, Berlin Heidelberg

Storbacka K, Nenonen S (2009) Customer relationships and the heterogeneity of firm performance. J Bus Ind Mark 24(5/6):360–372

Storey C, Kelly D (2001) Measuring the performance of new service development activities. Serv Ind J 21(2):71–90

Tatikonda M, Zeithaml V (2002) Managing the new service development process: multidisciplinary literature synthesis and directions for future research. In: Boone T, Ganeshane R (eds) New direction in supply chain management. Amacom

Vargo SL, Lusch RF (2004) Evolving to a new dominant logic for marketing. J Mark 68(1):1–17

Vargo SL, Lusch RF (2008) Service-dominant logic: continuing the evolution. J Acad Mark Sci 36(1):1–10

Verweire K, Revollo GE (2009) Sustaining competitive advantage through product innovation: how to achieve product leadership in service companies, research report. Vlerick Leuven Gent Management School

INNOVATION LAB
Electronic invoicing

MAINS Master, academic year 2007/2008
People and companies involved in the InnoLab:
Students: Tommaso Covino, Andrea Galavotti, Laura Sanna and Emanuele Taddei
Companies: Intesa Sanpaolo, Banca CR Firenze and SIA-SSB
Professors: Lino Cinquini, Riccardo Giannetti and Andrea Tenucci

1. The problem

The interest of firms in the theme of administrative process dematerialisation as well as the ensuing electronic invoicing has grown in recent years. Following the enactment of EU legislation[30] preventing public administrations from accepting invoices issued or transmitted in paper form, both in the public and private sector, realisation is mounting of the positive benefits deriving from the application of electronic invoicing models. Of particular note is the increasing quality in the delivery of public services and cost reduction by simplifying sales and purchase cycle management. It is estimated that savings can reach 2–3% of average turnover.[31]

The project aimed to understand the opportunities and criticalities of the application of electronic invoicing legislation to the administrative processes of the Scuola Superiore Sant'Anna.

In general terms, two types of electronic invoicing can be identified: e-invoicing in the strict sense when referring to all solutions designed to digitise and automate the process, which involves the creation of the invoice up to its filing. Instead, electronic invoicing in a broad sense refers to the integration and dematerialisation of the order-payment cycle, the analysis domain of which extends to the entire business and to administration logistics, from the creation of an order to the closure of the payment cycle and the related reconciliations.

It is in this second sense that electronic invoicing has the most interesting organisational implications. Electronic invoicing in this perspective is in fact the driving force behind the integration of the business process cycle.

The laboratory project aimed to analyse electronic invoicing in its meaning as an opportunity to rethink the management of the order-payment cycle in broader terms, with the aim of proposing an integrated and dematerialised model of the order cycle, identifying the potential savings resulting from the implementation of a technology platform for the integration of administrative, accounting and logistics processes.

2. Work methodology

The work was divided into two main phases:

1) The as-is model definition, as a representation of the process/service at the time of this study.
2) The development of the to-be model designed to represent the processes where the innovations required eliminate the criticalities of as-is.

Much of the work dedicated to the first phase concerned the definition of a method to represent the process in the most objective way possible in order to identify the critical issues and areas for improvement. This modelling work was conducted first through face-to-face interviews with the players involved in the process, thereafter transposing the modes of operation in graphic form into an inter-functional flowchart.

The team conducted interviews with players following a top-down approach, starting with senior management and subsequently more operational roles. All respondents, once the process was graphically represented, contributed to the validation of the data gathered to share their impressions on the correctness of the representation.

Based on the data thus collected, a measurement of the performance of process activities was implemented, focusing on the definition of the time and cost management of each activity. In the absence of objective process control, we relied on data provided during the interviews of those operating the activities. Where possible the "perceived time" was compared with quantitative data, using the dates affixed to orders or the document transition log in the management information system used for some parts of the process.

Time, as the object of the study, was divided into *value-added time*, where the operator carries out activities that are truly useful to the advancement of the process flow and *wait time* for each activity. Subsequently, the exceptional and reprocessing management activities were estimated. With the data obtained, an efficiency indicator[32] was calculated, defined as time spent on value-added activities on the overall throughput time. Considering the estimated hourly cost of the actors involved and of the consumables used, the team obtained a representation of the costs of different types of activities (costs of value added, wait time and exceptional activities).[33]

The table that summarises the data obtained with the method described above was defined as the *time effort worksheet* (TEW), representing the process activities as described in the inter-functional chart associated with the total time allocated to each phase (time), value-added time (effort), and costs incurred for each.

With the data thus organised for each activity and the representation of the process in graphic form, it was possible to identify the most critical phases and activities and hence the bottlenecks, the low value-added stages (unnecessary requests for authorisation, duplicated activities, correction of errors upstream, controls that are replaceable with greater delegation), the sequences with postponement of activities that generate critical information, excessive sequencing that increases the time required and the frequent occurrences of exceptions.

The model thus structured allowed calculation of an estimate of the average handling time of an order, from proposed purchasing to payment, evidencing that a large part of time management is composed of delays attributable to excessive sequencing of the processes and the teaching staff practice of asking to see the actual invoices.

The average order-management time was around 92 days, of which 79 are dedicated to wait time and 12 to reprocessing due to exceptions. The average efficiency indicator calculated on the entire process was particularly low and approximately equal to 0.7%. Regarding costs, the estimate stood at around €93 for each order, 70% of which are administrative staff costs.

For contract management (sales cycle), instead, a time of approximately 14 days was recorded, with a unit cost of €34 and average efficiency, calculated according to the indicator presented, of 1.8%.

The as-is process modelling, in addition to defining the inter-functional flow chart system and the definition of time and cost of the TEW table, is the basis on which the team then estimated the impact of changes in the processes as a result of the adoption of a multi-technology platform that would allow minimisation of costs and process times.

3. Solution proposed

In developing the to-be model, the team took into account the opportunities offered by a set of enabling technologies. Their strong interoperability and the automated process management provided numerous benefits.

The approach followed was aimed at implementing the subsequent paradigms:

1) Dematerialisation of documents and introduction of digital signatures to ensure that digital documents have the same legal status as signed paper in addition to the appropriate speed of information in real time. In this way, the amount of paper circulated would be reducedand it would be possible to maintain a single original copy of the document from which the entire life cycle could potentially be traced.
2) A multi-channel notification and authorisation system across multiple electronic devices, with the possibility of approving provisions and documents using "digital checkmarks", an optimal solution for reducing the impact of bottlenecks as well as wait time.
3) Full information sharing, as well as the uniformity of procedures to streamline processes, reduces the number of exceptions to favour inter-

action within the corporate perimeter and between companies and external stakeholders (suppliers, banks).
4) Service-oriented software and architecture paradigm (SOA) to simplify the integration of banking services within their application and to allow interaction with customer and supplier systems.

The inter-functional chart was then redesigned taking into account the as-is criticalities and the available enabling technologies, with the result that the proposed model reduced the 60 activities managed for the entire process to only 15.

With this greatly streamlined process design, the operating costs were estimated using the same as-is indicators to allow their comparison and the economic efficiency evaluation of a possible transition from as-is to to-be.

To estimate the crossover time of the new process, the activities were classified according to three levels of complexity that were assigned a hypothetical completion time (5 minutes for low-complexity activities, e.g., electronic signatures; 10 minutes for medium-complexity activities, e.g., receiving goods and delivery notes; 15 minutes for high-complexity activities).

The result obtained with the cost and time estimate of the to-be model saw a reduction of the cost of orders to €24 and the contract management cost to only €8, with an average estimated reduction of costs compared to the as-is process of over 70%.

A further stage of the comparison between the two situations was cost equivalence time estimation to determine the period over which the new proposed model begins to produce managerial savings compared to the initial situation. For this reason, fixed costs and variable costs were distinguished as well as calculating the weekly process management costs of as-is invoicing compared with the new solution.

The former came in at a cost of €9400 per week, while the latter carries a cost of around €2400 per week. The meeting point between the two cost curves identifies a cost equivalence time that corresponds to approximately 43 weeks.

The feasibility study that was presented highlighted the potential of the solutions to integrate and dematerialise the entire order cycle as an instrument to reduce operational costs, improve the accuracy of the process and significantly reduce the average implementation time of activities.

Part II
Innovative experiences in service management models

ICTs as a condition and as an enabling driver of service science in Italy

7

Guido M. Rey

The new international division of labour has changed the position of industrialised countries in world markets. The delocalisation of production processes and the dispersion of markets impose extensive use of ICTs (Information and Communication Technologies) to enable real-time presence where and when potential demand is expressed.

The technological leap to be made concerns not only production but also forecasting and managing demand as well as flexibility in relationships with suppliers and customers. ICTs, in this scenario, to express their full potential, impose an extensive change in firm strategy and management. A firm that uses a network to exchange information, knowledge and communications cannot limit itself to fragments of the process but must include all human, technological and financial resources in this change. This process cannot be left to individuals but must involve the network of relationships that characterise firms and markets. This chapter suggests a role of ICTs as enabling and decisive technologies for the development of an entire production sector until a new economic system is arrived at through successive steps and traditional production activities are abandoned.

7.1
Evolution of services and the Italian delay

In economic theory, too little attention has been paid to the service sector. Services are intangible assets that are incorporated in the enjoyment of a good and, in many cases, are implicit in the market concept and in exchange activities.

Many of these activities take place within the firm and, furthermore, whilst most services can be provided directly by the manufacturer and/or owner of the good, it

G.M. Rey (✉)
Istituto di Management, Scuola Superiore Sant'Anna, Pisa, Italy
e-mail: g.rey@sssup.it

L. Cinquini, A. Di Minin, R. Varaldo (eds.), *New Business Models and Value Creation: A Service Science Perspective.* Sxi 8, DOI 10.1007/978-88-470-2838-8_7, © Springer-Verlag Italia 2013

is a cost and efficiency evaluation that leads to choosing specialisation and hence favouring intermediation provided by professionally qualified organisations. This make-or-buy option can be modified by available technology and the advantages offered by activities that sustain the efficiency and efficacy of the core business. In sum, management must identify the phases of the production process that manifest the greatest capacity to create value and its appropriability.

Overall, the situation of Italian services is unsatisfactory, which is determined, on the one hand, by the poor competitiveness of a sector protected by the norms and characteristics of the output and, on the other, by the relevant diffusion of micro and small firms and the vast grey economy that has hindered the development of a large part of traditional services. There are examples of innovative firms, also in services, where information, knowledge and communication have assumed a strategic dimension and hence technological networks and relationships have developed. This development has led to the dismantling and relocation of many manufacturing processes (outsourcing, e-business), the actual and potential expansion of the market (electronic commerce) and the reengineering of procedures including firm governance. Service provision in technological networks has led to the development of platforms that encourage the encounter between supply and demand of goods and services, as well as relational transactions on both sides of the platform, and these multilateral services provide profits for the operators who use them and also for the platforms that provide them (Basalisco and Rey 2010).

The recent spread of service networks supported by numerous technological networks has signalled that in many cases the market analysis model is obsolete and should be upgraded. In fact, the sharing of fixed costs inherent in the processes of information transmission generates economies of scale, network and scope but also shifting costs, and these elements justify signing agreements between producers and the intervention of special authorities to ensure the benefit for consumers (Varian 1999).

In recent years, the services sector has continued to grow and in Italy its share exceeds 70% of total value added at current prices; in 2008 its surge seems to have come to an end if we exclude real estate activities (residential tenancies), which registered a sustained trend following the collapse of the real estate bubble, which also affected Italy.[1]

Without facing the theme of the evolution of the Italian manufacturing sector, one can state that in the last forty years, the agriculture and manufacturing industries have downsized in favour of the service sector (Table 7.1).

This transfer from manufacturing to business services has led to recognising and measuring the contribution of services previously often incorporated in the core business activities of enterprises. This phenomenon of outsourcing took on a notice-

---

[1] The revision of national accounting manuals (ISTAT 2004) has defined a tripartite division of sectors: the first aggregates and distinguishes traditional intermediation services; the second includes financial intermediation, the location of real capital and services destined for firms; and the third includes services offered by administrations, and services destined for people and for recreation. Linking these different services is the difficulty of assessing the value added at constant prices and productivity of resources used, while revenues, expenses and profits at current prices are easily quantifiable.

Table 7.1   Value added at factor cost (current prices) (ISTAT ESA 95 scheme)

Percentage composition	1970	1979	1992	2001	2008
Agriculture, forestry and fishing	8.9	6.5	3.6	2.9	2.3
1) Industry	30	30.9	24.5	22.8	20.6
of which: *Manufacturing*	27.5	28.9	21,8	20.3	18.1
3) Construction	9.2	6.9	6.2	5.3	6.2
4) Trade, hotels and restaurants, transport and communications	21	22.8	23.7	24.3	22.1
5) Monetary and financial brokerage, real estate and services for businesses	14.4	16.2	20.9	24.6	28
Private services (4+5)	35.3	39	44.5	49	50.1
Private services net of real estate activities			34.8	37.3	36.1
6) Other service activities	16.6	16.7	21.2	20.1	20.8
Total services	51.9	55.7	65.7	69.1	71
Total value added	100	100	100	100	100

able dimension in the late 1970s and 1980s, since both entrepreneurs and managers tended to focus corporate resources on core business to improve business efficiency and reduce the cost of services.

Amongst suggestions to promote the growth of the Italian economy is recourse to advanced services (AS) to improve competitiveness. Their scarcity is numerically verified by the input–output matrices that ISTAT elaborates every five years. A closer look shows that professional services are the most requested for reasons of specialisation and also because of the heavy administrative burden that Italian legislation levies on firms.

In manufacturing, the advanced services share of total intermediate purchases is between 5% and 8%, but professional services are nearly three times the percentage of IT and R&D services (Table 7.2).

The share of AS in the tertiary sector is much higher, peaking at 40% for AS within the same subsector, while in other sectors the share is around 15–18%.

Even in the case of the tertiary sector, professional activities account for between 60% and 80% of purchases of AS. The percentages of AS of total production value of individual subsectors vary between 10% for trade, etc., and 4.1% for financial intermediaries, real estate and business services. However, this subsector acquires intermediate purchases as 22% of production monetary value compared with nearly 50% percentages for trade, transport and ITC production (Table 7.3).

A recent survey on innovation in Italian firms (ISTAT 2010a) provides a clear and comprehensive picture of their behaviours. It was found that almost half of industrial firms innovated in the period 2006–2008, while for services the share drops to a quarter. For firms with 250 or more employees, as expected, the percentage rises to 80% and 55% respectively. Among the positive elements is increased product and process innovation, while previously this mainly concerned new products. Another

Table 7.2 Advanced services (AS) in manufacturing activities, year 2005

	Traditional manu-facturing	Chemicals	Metal products and minerals	Machineries and vehicles
Share of intermediate purchases on production	76.1	78.3	68.5	70.5
AS share on purchases of internal intermediary products of which:	5.0	5.1	7.3	8.1
Machinery leasing				
Computer and related services	0.3	0.3	0.5	0.4
Research and development (R&D)	0.5	0.7	0.8	1.2
Professional activities	0.3	0.6	0.2	0.5
AS/production value	4.0	3.6	5.8	5.9
Share of intermediate purchases on production				
AS share on purchases of internal intermediary products of which	3.8	4.0	5.0	5.7

Table 7.3 Advanced services (AS) service sectors, year 2005

	Trade, hotels and restaurants	Transport, postal service and telecom-munica-tions	Financial intermedi-ation and real estate	AS	Public sector and other services
Share of intermediate internal purchases on production	55.6	55.1	22.2	46.1	28.7
AS share of intermediary product purchases of which:	18.4	15.6	18.5	39.7	15.1
Machinery leasing					
Computer and related services	1.0	1.0	0.1	1.3	0.6
Research and development (R&D)	1.8	4.4	7.3	9.4	1.6
Professional activities	0.3	0.6	0.1	1.5	0.2
AS/production value	15.3	9.5	11.0	27.5	12.8
Share of intermediate internal purchases on production					
AS share of intermediary product purchases of which	10.2	8.6	4.1	18.3	4.3

element of interest is the link between technological innovation, and organisation and marketing in more than half of innovative firms.

In summary, the share of services in total value added is increasing, but AS play a minor role in this advancement and the innovation trend is not satisfactory.

7.2
The Italian delay in the diffusion of online network services provided by ICTs

Another critical element was identified in the insufficient growth of gross invest-ments in industry since 1992 (1.3% per year), resulting in a lack of innovation. In-novation design requires not only investments in hard technologies for production but also in strategic functions such as marketing, business finance and, especially, organisation, including governance. These innovations require investments in in-formation society technologies, but even for these, after an initial period of growth (6.4% per year between 1992 and 2001), development was negative (–4.2% per year between 2002 and 2008) and for the entire economy the respective annual rates of growth were 3.8% and 1.5%.

Italian firms use technologies, including online network services, to reduce direct and indirect operating costs, improve efficiency and get information on markets, prices and competitors but without substantially changing the relevant organisation, procedures and governance. Entrepreneurs believe these innovations lead to losses, which could be the case if the firm does not adapt its management systems to in-novation and if the information system is unable to accurately measure firm perfor-mance. Small and medium enterprises (SMEs) are particularly reluctant to use online network services to reduce transaction costs, improve market presence and offer a package of services related to the offer.

In fact, it is difficult to identify the contribution of ICT services to the growth of value-added trade because a number of factors interact within a complex organisa-tion. However, there is no doubt that in Italy, online network services have not been able to counteract the deflationary result that has rendered its business less compet-itive. The investments required to take advantage of online network services are not particularly large but require a mindfulness and willingness to exploit opportunities for change and improve corporate performance; it is thus vital to rethink and reengi-neer business processes. In awaiting the steady spread of online network services, Italian firms remain wary and wait for their sector or supply chain or customers to provide clear signals of their willingness to work on the network. Banks, however, are moving towards services involving firms and families in the diffusion of online banking services. Significant but partial initiatives have also been implemented in transport and tourism as well as in telecommunications. Public administration initia-tives are not lacking but, in general, their sites provide information and forms, while operability is only provided by a few administrations (Clementi and Rey 2010).

Investigations on the diffusion of ICT services (ISTAT 2010b) indicate a certain degree of reluctance in firms and families. Virtually all firms have these technologies, but there are still delays in Internet availability in small enterprises (10–19 employees). A comparison with other EU countries places Italy in an intermediate position but, overall, the delay is evident if a comparison is made with countries of a similar size and development.

Online network services for management are often on an intranet to protect the distribution of data, information and especially documents within a firm and/or industry. Intranet is available in, on average, three out of four large firms, approximately 30% in service sectors and around 20% in industry.

Almost all firms are connected to the Internet, with small differences by size and location. Among services, only banking operations are made online by almost all firms, using the Internet and interbank network. The technological solution adopted to favour online banking services was designed to enable real-time transactions without affecting the customer's organisation.

Two thirds of firms obtain information on the Internet while only half use it to receive services from suppliers. Regional differences are modest, while the differences due to firm size are significant. However, there is room for improvement in some services for large firms, too, such as personnel training and after-sales service.

For over 60% of firms, the website is a fundamental tool enabling dialogue with the outside world, applicable to over 90% of large enterprises. For over a third of service firms the website is a tool to make reservations and receive purchase orders, in other cases, albeit less numerous, it is only a showcase or a sophisticated virtual store.

To improve processes and reduce inter-transaction costs, firms use automated data exchange in just over a third of respondents, a percentage that doubles for large firms. It is interesting to note that the distribution of services within this subset is broadly similar, irrespective of firm size. The percentage for exchanging information on orders and products is over 60%. A considerably popular service in recent years is electronic invoicing, which for the time being sees a clear prevalence of inbound electronic invoices (approximately 80% for small businesses) compared to outbound invoices (35%). This data confirms that large firms make greater use of online network services, especially firms delivering energy and telecommunication services.

Commenting on data information exchange with suppliers and/or customers in relation to supply chain management, it should be remembered that the results depend on the spread of services on both sides of the network. For example, large suppliers/customers may require this tool to enable dialogue with their counterparts if expecting a reduction in transaction costs or an improvement in the definition of orders and/or control production timing and/or logistics. For the time being, this is a limited practice that involves only one-fifth of firms, a figure that rises to one-third for large enterprises. In this subset of enterprises, the distribution of the items does not change with regard to firm size and sector; in particular, information on stock situation and demand forecast concerns over 70% of the firms in question while this percentage rises to over 80% for information on deliveries.

It has often been observed that the inadequate diffusion of e-commerce in Italy is attributed to a number of reasons ranging from the customer's need to physically see a product before making a purchase, to the risk concerning the transaction outcome. Doubts remain regarding the legal protection of the parties involved and, also in this case, the main conditioning factor is the absence of network counterparties that justify investment in electronic commerce.

These elements have a lesser impact in terms of online purchases made by firms with regular suppliers because the exchanges are more frequent, the object of supply is defined and the larger firm often can impose the use of online transactions. This aspect may explain the prevalence of online purchasing for the tertiary sector, 30.8%, against 26.4% for the industrial sector. These percentages increase with firm size but, nevertheless, the value of online purchases does not exceed 5% of total purchases, irrespective of size and location.

To understand the degree of organisational innovation observed in the case of online purchasing, over 40% of firms involved share relevant internal information such as accounting and inventory management. Although the percentages do not vary between industries, services and location, they do increase with size (doubling from small to large enterprises).

The critical element for online trade is sales, since only 4.8% of firms use this channel, with a clear difference between industry (2.3%) and services (8.5%) for an amount that, in 80% of cases, does not exceed 1% of total sales. In terms of sales, size plays an important role, moving from 4.5% for small business to 13.7% for large firms. These figures deserve consideration in terms of the use of online services, particularly in the relationship between firms and consumers.

It is interesting to note that for firms that sell online, the percentage of internal information sharing is very similar to that reported for purchases but evidently applied in much smaller proportions.

More advanced firms can be disaggregated according to whether the technologies are used mainly to improve management efficiency (of the firm and the supply chain) – the majority – or they are designed to exploit market opportunities and increase competitiveness. E-invoicing would allow "linking" the two "instances" of the integration process (internal) and exchange (external) since it is positively associated with the spread of online debit and payment services.

In conclusion, innovation efforts through ICT progress slowly and are mainly concentrated on administration and finance, as occurred previously in the earlier ICT evolution phases. The recent awareness that growth calls for an effort to improve the dissemination of information and knowledge imposes an effort to clarify the benefits of the suggested strategy.

SMEs have a low propensity to use technological networks to develop online relationships. The reasons for this resistance can be manifold and related. Entrepreneurs' knowledge of these technologies may be lacking, but the same could occur to their employees and associates. Furthermore, the market's capacity to provide clear and unambiguous signals favourable to these network services may be considered insufficient. Additional barriers are: (1) lack of a structured and shared model to implement online transactions; (2) difficulties in the management of heterogeneous organisational schemas; and last but by no means least (3) fear that online transactions leave traces that may be used by the revenue authorities.

7.3
Online network services for the networked firm

The recent debate on the causes of inadequate Italian economic growth established the convergence of at least three aspects: (1) the need for innovation to improve the competitiveness of products and services, (2) the insufficient endowment of tangible and intangible externalities and (3) the necessity for fiscal consolidation and reduction of public debt (Ciocca and Rey 2004).

The first element indicates scarce attention to new technologies and to the contribution that market knowledge and streamlining of logistics can make to value creation. The second element indicates that an economic system evolves along with institutions and the changes must involve both. The third element suggests dual actions to improve the delivery of public services at equal cost and contribute to the increase in real GDP without increasing public spending, if anything, reducing tax and social contribution rates.

The innovation gap that is still observed in many service firms is likely to continue to distort strategies, both private and public, if not actively innovation driven, including firm demand for AS, and if competition does not increase together with the reduction of the grey economy. In an increasingly integrated economy, a stable competitive advantage for Italian firms should be the ability to provide customers with higher value added goods and services. This competitive advantage requires an entrepreneurial ability to develop strategic and managerial components in the face of both product/service design and marketing. Innovation in products/services thus requires the development of processes based on knowledge management, where the more traditional skills are complemented by strategic business analyses. Information, knowledge and communication, for example, change human resource recruitment, revise ICT management and promote integrated logistics. These are just some examples, but methods for sharing knowledge in the relationship between suppliers and customers are also important (Nooteboom 1999, 2000; Gulati 1998).

Lacking a common vision on the role of business services, within which government services, the individual company or industry group will continue to rely purely on technological innovation to improve its competitiveness since they rightly consider incapable of conditioning the offer of the service sector on their own where many influential guilds, alongside some large conglomerates, exploit their market power. This passive approach entrusts the solution of problems to public policy but, unfortunately, there are insufficient resources to encourage innovation and to invest in infrastructures. Only a cognisant and proactive attitude of entrepreneurs and top managers can contribute to the success of a crisis exit strategy.

Prior to suggesting innovations, it would be appropriate to verify how information and knowledge are distributed over the network between the firms concerned and what effects this can have on creating innovative services (Shapiro and Varian 1999).

Relational networks can be interpreted as a set of transactions between firms in order to coordinate their business functions without limiting competition and represent a source of efficiency compared to other forms of production and exchange organisation (Gulati et al. 2005). In this context, the coordination of transactions is an intermediate solution between the services provided by the market and the integration/merger of two or more firms, using network technology to link the participants in the initiative (Williamson, 1999; Baker et al. 2008).

However, technological networks are limited in their capacity to integrate and disseminate enterprise information, while it is essential for relational networks to assess: (1) if participating firms are complementary and/or specialised, (2) the characteristics of the available resources in internal processes and (3) the sharing and exchange of information and knowledge in either a structured or open form (Sallusti 2010). Really important is selecting the development path of relations taking care of the difference between expected benefits and costs and the efficiency in the performance of different activities implemented with the aid of networks. Finally, the success of relational networks requires careful selection of the governance model to be adopted since an entrepreneur is unlikely to relinquish his power, otherwise sharing only part of it with other entrepreneurs, in the absence of obtaining compensation in the network governance.

The relational schemas can differ depending on the capacity to generate knowledge and can vary with localisation distance, available technologies, coordination and organisation design. The variability of schemas is associated to resistance demonstrated by small business owners, thus suggesting that a strategy suited to Italian SMEs requires flexibility and adaptability if intended to foster widespread innovation that improves the competitive position of Italian firms.

Problems can arise when measuring the expected value of the relational network and/or selected services, i.e., giving an economic value to the benefits and costs of the expected change (see Chap. 6).

Similar problems arise when the transaction is between public administrations and citizens/firms to provide services. To avoid the charge of inefficiency, public administrations and monopolistic enterprises apply parameters for the calculation of direct and indirect costs similar to those provided by the market, including queue time. Governmental accounting has continued to be the point of reference for some services reserved for public administrations since they are not directly usable by the individual and the monetary counterpart is in the form of taxes. Under monopolistic conditions, whether public or private, incentives to change collide with the resistance of conservatives who, while not perceiving their advantage, clearly perceive the risk implicit in the new business organisation.

An evaluation of the network relations is provided by the definition (real or virtual) of transfer prices within the large enterprises, since this dictates the share of the value created by the company's organisational unit. In this case, the potential conflict is resolved (not always) by top management but the consequences can be severe if the solution adopted discourages collaborative behaviour of each division (the issue of transaction costs conceptually associated with information asymmetry in a complex business system).

In the event that firms used the network limited to technological services, costs are defined and monitored by the specific authority; in case relations generate information and knowledge exchanges that are considered as equal by the parties involved, then the problem does not subsist. Likewise, the problem does not arise when relationships streamline activities that were previously available, or had a market price or internal budgets that allowed valorisation of the variant envisaged. On the other hand, if the relations are actually innovative and involve the core business or, rather, the functions that generate their strategic value, the cost–benefit analysis becomes complex and necessarily subject to negotiation between the business partners, although arriving at a meeting point is unlikely in the absence of undisputed leadership. The possible solutions are, therefore, the market or the acquisition/merger of firms participating in the exchange of knowledge. These alternatives can explain the resistance of enterprises, especially SMEs, in relation to networks and online network services and their decision to purchase technological services or services provided by specialised firms, since the transaction ends with its payment and the relational exchange is limited to information that is not always structured. In this situation, for the deployment of online network services, the professionalism of decision makers is crucial, as is the trust and credibility of partners. The possibility to assess the risk in its various components and its insurance premium is also relevant. In relational networks it is vital to clarify the potential objectives and the governance rules, languages, technologies, distribution of benefits and costs, and their measurement. These factors recall the second point that emphasised the need for the involvement of both firms and institutions.

7.4
Online services for SMEs and the role of large firms, banks and government

Surveys on the diffusion of ICT in Italian firms have verified that online services are mostly used by large firms and the percentage tends to decrease with firm size (see Sect. 7.2). Statistics on the size of Italian firms show that 98.1% of enterprises are small and micro firms engaging 58.7% of employees, while firms with over 250 employees number 3508 (0.08% of firms), employing 3,214,387 workers (18.7% of employees). Most importantly, the share of value added trade of the former is 44.1% of the total while the share for large firms is 28.7%. These findings can explain the delay suffered by Italian business and call for suggestions on the strategy and development envisaged for the whole economy. Not all industry branches are as fragmented as the construction industry and the private service; industries that adopt technologies characterized by economies of scale (durable goods, utilities, banks, etc.) or are trading in foreign markets have higher average sizes. For the completeness of the description, it should be added that about a third of SMEs are part of either an Italian or foreign group.

In respect of the few large firms, there are several hundred districts where small businesses predominate and where very few medium-sized enterprises exert their leadership. Externalities are generated by the agglomeration of activities and by the wide range of products in the same industry offered by SMEs in a defined territory. Externalities are such with respect to the individual firm, but are certainly internal to the district and even characterise it. Also belonging to districts are activities that are not strictly industrial such as ancillary services and technological, relational or specialised industry networks that are not all related to the location but to the various phases of the production process. In districts, firms operate with a collective vision, even if not formalised, so that innovation spreads by imitation and is coordinated by the manufacturing process and the market. This endogenous diffusion is difficult to reconcile with network services as required by the internationalisation processes and as suggested by the use of networks to integrate production, commercial, administrative and financial activities.

Essential innovations require technologies and services as well as relational networks to share knowledge and take advantage of complementary competencies. SMEs are facilitated in their decisions by the downsizing of technologies and the specific service architectures in cooperative systems (service-oriented architecture (SOA), see Jones 2005). Downsizing is not only a technological opportunity but should encourage consultancy and knowledge services to intervene, even through a few representative district firms, provided that districts are willing to take the path suggested by service science, management and engineering (SSME) towards integrated services.

Innovation passes also through interaction between industry and services and/or between large and medium enterprises on the one hand and small and micro enterprises on the other. Innovation is driven by competition and by technologically updated enterprises and does not necessarily require an increase in the size of firms involved in modernisation but certainly requires high-level and professional training for managers and employees.

Often legislation discourages innovation if lawful authorisation is required; a similar disincentive can be created by public administrations and rules if there is a delay in their use of the network services. In Italy this problem has been addressed and resolved in most of the large public administrations as the legislation is adequate (Rey and Clementi 2010).

The list of available network technologies is long but not all are essential or suitable for SMEs. Investigations show that the few firms using network service technologies focus mainly on e-business and in particular on administrative and financial services, while less common are services related to demand planning, CRM and e-commerce in general. Electronic invoicing is spreading although not always consistent and coherent with the straight-through processing (STP) methodology.[2]

The service sector, although delayed, has given rise to innovative multi-professional interests, as appropriate to improve services and reduce costs, eliminate repetitive or labour intensive tasks and integrate production technologies with technology management. Combinations of technologies have been implemented and used to innovate procedures and then adapted to different applications and services (e-gov, e-health, e-banking, e-etc.).

---

[2] In this chapter, attention was focused on firm and production system integrated services, but it also explores SSME integrated public services and e-gov (Viscusi et al. 2010).

To reduce costs and keep applications up to date, the network nodes can make use of service providers and build platforms designed to perform transactions to implement the green economy by reducing energy consumption of work-related travel.

In sum, technologies are available to timely update network clients while the relative cost can be sustained on the basis of the user's needs and paid according to consumption.

At this point, the question returns: Why are micro and small businesses reluctant to use technology services? Some answers were provided previously, recalling the requirement that the network architecture must foresee the possibility of some missing or inactive nodes and solutions are available to bypass the critical points. In this way, the network can also begin to operate in segments without waiting for a complete list of participants and applications. Networks should allay fears on the security of transactions but it is essential that physical and logical safety is guaranteed at all points of access, even within firms.

Barriers to the spread of technological services also derive from the difficulty of meeting demand from SMEs when service providers limit themselves to supply technologies. It is necessary to find solutions that aggregate potential users, but this process of diffusion by imitation should occur independently, since time, in this case, is not an ally. To overcome these barriers requires the innovation project to be shared by SMEs and developed more effectively with business leaders from the district or local sectors, without public administrations necessarily acting as intermediaries.

Networks and technology services will only spread if both national and local production systems participate – at different times and in different ways – in commodity and financial network transactions between firms including SMEs. The decision to introduce these innovation processes (online transactions) and products (goods and services online) or both cannot be left exclusively to the existing initiatives since the network economy produces substantial benefits only enhancing users and available services (network economies).

Three groups of actors must contribute to the development: (1) firms providing technology services, (2) large firms that find expediency in the implementation, through the network, of relational and commercial transactions with other firms to increase knowledge and/or with families to reduce costs, and (3) banks to facilitate the settlement of monetary transactions.

This tripartite division, on one hand, may be an opportunity to share the strategy, but on the other hand could be a restraint because the differences in local conditions may induce large national operators to delay implementation of the part of the project that concerns them and/or is in their jurisdiction.

Network service providers must have demand expectations in a competitive environment that push towards efficiency and the creation of technological infrastructures without trying to exploit their technological monopolistic position and thus adopt open solutions. Large firms should encourage their suppliers and customers to face tangible and intangible investments in ICT, in the mutual belief that increased efficiency leads to increased turnover for both large firms and SME suppliers of goods and services. Indeed, large firms can provide this expectation of demand for online transactions justifying the investment for SMEs. Public service providers (ENEL, ENI, Telecom) could develop these initiatives aimed at increasing the efficiency of the Italian system since they are in contact with all Italian firms and can obtain undoubted advantages from an innovation project in the management of relations with customers and suppliers.

Banks have different tasks and roles: (1) the online receipt and payment system is already substantially widespread, due to the security conditions that it guarantees and the modest organisational changes required, and (2) the banking system has considerable experience in online services and transactions and has already walked the path of organisational changes resulting from the use of online services. Besides providing the experience of secure solutions, banks can provide guarantees for the counterparty in the transaction, acting as a "third party".

An important but external role is played by the public sector since: (1) e-government leads to the use of the network in contacts between firms and administrations (tax, social security institutions, chambers of commerce), (2) public investments in infrastructures create externalities functional to the development of online services, (3) public demand accounts for 17% of total demand and (4) investments in computer and organisational training reduce the professional gap that exists, especially among entrepreneurs and SME employees. In the case of public administrations, the territorial design is an easy reference for public services since these are, by definition, geographically distributed.

The different types of businesses can also be grouped into associations, consortia and districts according to the production chain and/or territorial sphere, etc., and thus some working hypotheses relating to the firm can be extended to these collective actors.

7.5
Suggestions for alternative innovation strategies

It would be incorrect and misleading to wait for industrial policy to enable an innovation strategy through service networks; its role is limited to the provision of infrastructures and incentivising initiatives.

Innovations are a business decision but also banks are involved in financing the innovations. Only the cognisant and proactive attitude of entrepreneurs and senior management can contribute to the success of a large firm's strategy to exit the crisis, taking with them the small and medium-sized suppliers of intermediate goods and services. It is for top management and the corporate control group to define this overall strategy that aims to change not only the tangible (products and processes) but also intangible components (organisation, knowledge sharing, rules) as well as its relations with the outside world, without losing the strengths of the firm and in particular the knowledge wealth built up from experience, information assets and professional human resources.

Involving relational networks can be onerous for the enterprise due to the initial costs as well as the potential loss of intangible assets available to the firm, but the expected benefits are greater because the firm's efficiency improves and a positive effect is transferred to the business environment.

These strategies require not only a selection of products that lend themselves to their transformation into a network service but also the organisational and relational consequences involved in change, as well as the immediate or potential availability of skilled human resources together with adequate funding.

Managing change has a medium to long time-horizon and requires a correlated evolutionary vision, but the individual parts that make up both the innovation process and the re-engineering of procedures must be defined according to a schedule that provides for their quick launch.

Further examples of innovation consist in replacing the sale or rental of the product with supplying the entire service associated with the product to meet customer needs and increasing the value of the offer for both the manufacturer and the customer (e.g., photocopier supplier managing and archiving documents). Firms involved in this change can be manufacturing firms selling integrated services or tertiary firms that acquire and then integrate services or, finally, firms that coordinate the link between products and turnkey services. From this simple list it can be deduced that innovation can make it difficult to assign a firm to an industry rather than a service. Firms that coordinate services do not necessarily have to be large but must be highly equipped with technological and relational networks as well as professional and financial resources, and demonstrate to customers and lenders their growth potential and their credibility as reliable partners.

An example of success is provided particularly by medium–large firms that have established themselves on world markets through the extensive and innovative use of ICTs. To achieve these results, they have relied on the diffusion of knowledge within the firm, the relocation of more simple and labour intensive activities, and the selection and acquisition of outsourced services as well as collaboration with other innovative firms (R&D marketing, logistics, etc.). In sum, many of the service lines included in SSME have been successfully adopted (Katzan 2008).

The Italian production structure and its localisation suggest launching coordinated initiatives at the local level to promote the diffusion of technology services.

The main objective to be achieved is strengthening the planning capabilities of the system of enterprises, institutions and local banks to develop strategies and projects fostering the use of new technologies, together with promoting research, training and the adoption of network standards. Thus local initiatives (for example, in districts) must involve business associations, public institutions, banks and other local organisations (chambers of commerce).

The strategy suggested (Varaldo et al. 2007) assigns to demand the definition of priorities and actions of promoters and service aggregators to support their spread. The governance should be based on shared objectives and risks, since a network of services has high initial costs and risks such that no small or even medium firm is willing to undersign a reorganisation and investment plan without guarantees of control over the management of the initiative and the project's consistency with the expectations of its potential suppliers and customers as well as competitors. Dissemination requires interaction between suppliers and customers to understand their needs and reduce potential waste due to the continuous adaptation of the applications as a consequence of the increase in services and users.

At the start, "trigger nodes" in the network need to be identified to create an "epidemic type" diffusion process as well as competitive advantages for participants. Reliable, appropriate and affordable technological and organisational solutions are available but there is also a risk of increased servicing and maintenance costs since a network firm cannot come to a halt as a result of the absence of a connection or an application bug.

Beyond integrated business services, integrated production systems (e.g. production chain, districts, automotive, local production systems) are also among the list of services available to the network in different institutional settings.

A sector and/or a territory sector with a large number of operators has considerable coordination difficulties together with the risk of opportunistic behaviour caused by the inevitable spill-over effect generated by investments in infrastructures. It is not just the amount of participants that renders a solution problematic; it is also influenced by the quality of the operators involved.

An alternative strategy suggests the use of a technology platform to help SMEs overcome the technological and cultural barriers to product and process innovation and to accept the necessary organisational adjustments.

There are several platform models for firms, sectors and markets, depending on the contribution that these platforms provide to efficiency and value creation for those firms participating in the initiative (Basalisco and Rey 2010).

The simplest model assigns to the platform the role of intermediary in transactions between two groups of members working with the respective organisations; this is the case, for example, in online banking services. The platform creates value in the form of cost reduction and operators can measure not only the economic but also the relational dimension (model A).

In the second model, the platform offers both technology and market coordination as well as providing a powerful impetus to the organisations that control it. In turn, participants are encouraged to develop new markets linked to products and/or services that are complimentary to the transactions foreseen on the platform. The development of these innovations also requires technical rules suited to limit *legacy*. An example is the platform implemented by firms in a sector through which common projects, research or supply of intermediate inputs, are attained (model B).

The third model describes the situation of a firm that has achieved a competitive advantage from the exploitation of a patent or an innovative process and seeks to consolidate its lead with input from suppliers of services that are complementary to the original innovation and also from customers who use the product/service. Extended application solutions are obtained that increase innovation, and widen the spread and number of participants to the platform. Notable examples of achieving externalities associated with large firms are Cisco, Intel and Microsoft (model C) (Gawer and Cusumano 2002).

The study of platforms has driven research towards a new model of economic analysis: economic models of bilateral markets where platform revenues are derived from two distinct groups of operators and depend on the interaction of different types of members of the platform. The business that the technological platform makes available will be successful and achieve profits only by creating value for the users of the platform and by distributing some of the benefits arising from the externalities created by the platform.

The number of potential users and the difficulty of persuading them to accept innovative technology services require firms to perform functions to promote and develop the business plan for new services to avoid the platform being reduced to a supply chain where the direction of services is one-sided. The promoters must establish the rules and organisational constraints to give credibility to their efforts to support and encourage the industry or the district and, in particular, foster participation of firms of all sizes to increase the number of services provided by the network. SMEs, to protect their competitiveness, would be obliged to invest in the network (and to change the organisation) to take advantage of opportunities created by a market where complementary services to the platform are exchanged and by the impetus that the network will gain from pre-existing business and/or geographical relations.

Complementary to the action of promoters is the role of aggregators, i.e., firms establishing the platforms and facilitating the work of entrepreneurs involved in the project, since the aggregators, in agreement with promoters, acquire technologies and provide services that are paid for on consumption and therefore do not require expensive investments in technology by SMEs.

7.6
Which policies for the development of online network services?

It is in the interest of Italian firms to proceed in this direction, in compliance with EU legislation, valorising the interests of their businesses on domestic and international markets. Within the EU, horizontal policies are also expected to improve relations between individual firms within the country and within the EU. This plan foresees full adoption of policies that encourage the spread of ICT among Italian firms, since this initiative will benefit not only domestic firms but also EU partners.

Once the strategy and related projects have been developed, the next step is the involvement of political authorities for their support and the definition of their interventions. Public policy must not destroy the existing equilibrium but must activate changes to accommodate developments in technology and markets. Within this institutional context, the technological platforms, with the variety of solutions available, promote the success of policies intended to help marginal firms strengthen their economic and cultural interests, their characteristics and their capabilities.

Instruments for the implementation of this strategy are numerous, beginning from public demand to foster the creation of ICT service firms in areas where SMEs are located, avoiding a concentration of ICT service firms in only the most developed areas of Italy, as at present.

Although there is legislation to encourage innovation as well as tax incentives for tangibles but also intangible investments, the introduction of methods to evaluate the effectiveness of these investments is essential.

The novelty of this strategy is the allocation of incentives only to the aggregator organism (the platform), which coordinates and supervises the implementation of the project, but no incentives for individual firms, who have a central role in the aggregator's governance. This policy recommendation follows an indication that the project does not only have management relevance, but is part of the sector's innovation strategy. It is the social advantage that should be encouraged while the individual entrepreneur must be induced to participate to obtain economic and managerial benefits resulting from the dissemination and utilisation of services.

Financial incentives may have negative side effects related to adverse selection if the proper tools to identify the best projects are not available, and can also allow unethical behaviour where an inefficient system of controls is unable to compel them to correctly use the funding obtained. It is essential to prevent the private sector from adopting a passive attitude and being only concerned with securing funding and incentives from the state, without taking responsibility for the implementation of the project. Unfortunately, the confusion in responsibilities at various levels of government slows down implementation time and activates resistance that shields conservative positions.

To avoid opportunistic behaviour, entrepreneurs and managers must be convinced that in case of non-participation in the initial phase, no advantages will subsequently be obtained. No less important is the role of promoters that can defend innovation against the market power of technology providers; dangers are inherent pseudo-consultants exploiting the digital illiteracy of many entrepreneurs and the difficulty of correctly measuring the benefits obtained from the services.

These promotional tools are noted and have been widely used in the past. The novelty is the priority assigned to network technologies that do not require large money assets since individual entrepreneurs do not have to acquire sophisticated technology, but obtain and pay for services on consumption.

ICT services are numerous and flexible so there is room for growth for SMEs where there is an expectation of market development that urges efficiency and thwarts rent positions.

By definition, the effectiveness of this policy instrument depends on cultural factors, since it focuses on knowledge and on the relationships between firms, but entrepreneurs should perceive new technologies as an opportunity to improve business.

Large firms that have so far focused their investments on technologies to improve production processes and have limited their investments to strengthen internal services can instead encourage the spread of networked firms by enhancing the business-to-business sector and exploit their market power to encourage suppliers and customers to perform online transactions.

It was previously suggested that potential promoters could be large enterprises and banks, both directly and through ad hoc firms, which can play a strategic role towards their customers and suppliers. However, their role could be much greater if they participated in selecting and financing both promoter and aggregator firms, in agreement with enterprises and local governments.

7.7
Conclusions

Surveys have revealed that in Italy the diffusion of ICTs among firms is belated, especially in medium and small firms. The service sector, although embedded in input–output activity, is not particularly advanced. Finally, firms operating in foreign markets show that network technologies are a key attribute in acquiring contacts with enterprises within and outside the EU.

The strategy centred on technological innovation and the enhancement of services aimed at defending Italy's position within and outside the EU requires that at all levels, territorial and dimensional, firms participate by using networks both in economic and relational transactions. Network economics precisely studies the benefits that can be obtained from increasing users and the variety of services available. Coordination difficulties are notable when the number of operators is high and a risk subsists that opportunistic behaviour is induced by the inevitable spillover associated with investments in infrastructures.

The Italian manufacturing industry suggests research on new architecture that is functional for territorial or production agglomerations, especially for leader firms in districts and/or for specific industry sub-branches. It is important to have ICT-based services and to show evidence of the benefits when operating both within and outside the network. Services provided by a technology platform can help SMEs develop transactions and exchange of knowledge with other firms, thus overcoming the cultural barriers and professional gaps, as well as implementing the necessary organisational adjustments. The tools available are manifold, beginning with public demand to encourage ICT service firms in southern Italian. In conclusion, we can state that the objectives are clear and policy instruments aimed at both the efficiency and competitiveness of the business are available, but its feasibility is not a foregone conclusion without the contribution of large firms and banks since the dwarfism of our firms fosters an attitude that worsens the delay in the dissemination of innovative technologies.

ICTs have passed the stage of simple data processing, distributed computing facilities and simple networking technologies. At present ICT networks provide application services and knowledge sharing, but unfortunately, in this pursuit, Italian e-business has been delayed and requires accelerated investments and development of applications designed and implemented for SMEs.

Assuming that this pursuit will be carried out independently by millions of small entrepreneurs is illusory, since none of them are able to impose dialogue and select the appropriate technology. Decisive solutions should be designed within the private sector and government can only intervene to facilitate their implementation.

References

Baker GP, Gibbons R, Murphy KJ (2008) Strategic alliances: bridges between islands of conscious power. J Jpn Int Econ 22:146–163

Basalisco B, Rey GM (2010) Enhancing the networked enterprise for SMEs: a service platforms-oriented industrial policy. Working paper, ArtDeco, Pisa

Berners-Lee T., (2000) Weaving the Web, HarperCollins

Ciocca P, Rey GM (2004) Per la crescita dell'economia italiana. Economia Italiana 2

Clementi S, Rey GM (2010) eGovernment initiatives in Italy. In: Viscusi G, Batini C, Mecella M (eds) Information systems for e-government: a quality of service perspective. Springer, Berlin

Gawer A, Cusumano MA (2002) Platform partnership. Harvard Business School Press, Boston

Gulati R (1998) Alliances and networks. Strat Manag J 19(4):293–317

Gulati R, Lawrence PR, Puranam P (2005) Adaptation in vertical relationships: beyond incentive conflict. Strat Manag J 26:415–440

ISTAT (2004) Metodologia di stima degli aggregati di contabilità nazionale a prezzi correnti. ISTAT, Rome

ISTAT (2010a) L'innovazione nelle imprese italiane (2006–2008). Statistiche in breve, December. ISTAT, Rome

ISTAT (2010b) Le tecnologie dell'informazione e della comunicazione nelle imprese. Statistiche in breve, December. ISTAT, Rome

Jones S (2005) Towards an acceptable definition of service. IEEE Softw 22(3):87–93

Katzan H Jr (2008) Service science. iUniverse, New York

Nooteboom B (1999) Inter-firm alliance. Routledge, London

Nooteboom B (2000) Learning by interaction: absorptive capacity, cognitive distance, and governance. J Manag Gov 4:69–92

Sallusti F (2010) Relazioni fra imprese e reti: analisi e studi di caso. L'Industria 1:83–113

Shapiro C, Varian HR (1999) Information rules: le regole dell'economia dell'informazione. ETAS, Milano

Varaldo R, Rey GM, Ancilotti P, Frey M (eds) (2007) La diffusione dei servizi innovativi in rete: linee strategiche (mimeo). Scuola Superiore Sant'Anna, Pisa

Varian H (1999) Intermediate microeconomics: a modern approach. W.W. Norton & Co. New York

Viscusi G, Batini C, Mecella M (eds) (2010) Information systems for e-government: a quality of service perspective. Springer, Berlin

Williamson OE (1999) Strategy research: governance and competence perspectives. Strat Manag J 20:1087–1108

INNOVATION LAB
Work organisation hypotheses based on Web 3.0 and Web 4.0 models

MAINS Master, academic year 2009/2010
People and companies involved in the InnoLab:
Students: Antonino Bordonaro, Paolo Bortone, Giacomo Carollo and Alice Guidi
Companies: Centro Ricerche Fiat, Intesa Sanpaolo and Poste Italiane
Professors: Fabio Baroncelli, Stefano Fontanelli, Mario Rapaccini and Andrea Tenucci

1. The problem
What will the Web of the future be like? The answer is somewhat complicated since corporate experience as well as the literature on the subject is undoubtedly wide-ranging, especially given the innovativeness of the subject. To date, the Internet has 1.3 billion users connected all over the world. In 2006, around 161 EB of information was created and by 2010, this traffic will have grown more than six-fold, reaching the gigantic figure of 1 ZB of information, which will multiply every 72 hours. The evolution of technology is revolutionising all interaction paradigms: the challenge is to move from the social Web (2.0) to 4.0, or rather, the Web of Things, where everything will be interconnected. With the "Internet of Things", all subjects communicate with each other and over 4 billion devices will soon be networked. Consumers will entirely depend on network services and new business models need to be rethought from a semantic Web 4.0 perspective, thus transforming the Internet into a key economic tool for firms that use it. However, as long as the Internet consists of unstructured data and resources, most of these will not be fully usable or exploitable by either firms or users; this requires the definition of conceptual maps that render data interoperable and able to interact.

Enterprise 2.0 is by now a reality: information from blogs, social networks, forums and wikis is a valuable database of knowledge for semantic search engines. Firms, leveraging on the evolution of Internet services, can thus identify customer needs and offer highly customised products and services. However, all this must be reflected in a level of trust in the confidentiality of sensitive data. Everything, everywhere and always – the Web of the future will be a 24-hour and 7-day service. Chat, news and e-mail will all be used by firms to communicate with their customers. With the semantic Web, all systems will be integrated and no data will remain unused. Thanks to the ontologies of structured conceptual maps, which allow the formal and explicit representation of a domain, it will be possible to create a global network of concepts linked by increasingly strong relationships.

The new Internet will allow safe and efficient access to network services from any device. These services will be interoperable, coherent, consistent, scalable and reliable. This is the Web of the future, the service Web.

How has the Web developed? The birth of the first website (Web 1.0) dates back to 1991; all static sites that serve as showcases form part of this category. Web 2.0 started creating interest in the early 2000s. This is the Web that we all know; it is more interactive gave birth to social network sites (Myspace, Facebook, blogs, wikis and YouTube). Today, the evolution of the Web of the future, namely Web 3.0 and 4.0, is already a topic of conversation, although the boundaries are rather fuzzy: the former assumes more of a semantic Web connotation while the latter considers the more futuristic solutions of the Web of Things. The evolution should be envisaged on a temporal axis: Web 2.0 does not rule out the presence of Web 3.0 and so forth.

The realisation of the Web of the future requires an understanding of the concept of semantic search, where the user can undertake targeted interrogations expressed in natural language: the machine, through cognitive and conceptual maps, will be able to attribute meaning to the questions posed. This type of search is beyond current technology; in fact, traditional search engines make use of keywords and cannot in any way interpret the context.

The creator of the Web, Berners-Lee, first offered a concrete concept of what the future of the Web will be: "The first step is putting data on the Web in a form that machines can naturally understand, or converting it to that form. This creates what I call a Semantic Web – a web of data that can be processed directly or indirectly by machines" (Berners-Lee 2000).

2. Work methodology
The team's work was developed over three key stages. The first was dedicated to understanding Web 3.0 and 4.0, which was followed by a scouting phase to explore emerging trends in the use of semantics within business processes. In the third phase, the workshop identified and investigated, as output, two very different but extremely interesting processes with a wealth of perspectives: technical assistance for Poste Italiane and a start-up initiative for Intesa San Paolo.

Once the literature review was completed, the group proceeded with the analysis of cases of applications of 2.0, 3.0 and 4.0 technologies in firms.

From scouting around 30 national and international firms, the team found that this technology is mainly applied in internal processes and operations, in particular in R&D, marketing, CRM and HR. The new challenge for future organisations will be the ability to consolidate these into a single platform where data is interconnected. This will make an immense knowledge management tool available to firms.

Together with the partner companies, the team decided to address two process organisation hypotheses with significant impact on the use of this technology: technical assistance at Poste Italiane (highly structured support process where large amounts of data have to be managed) and the Intesa San Paolo start-up initiative (a business activity that will increase over the coming years where information is heterogeneous and difficult to organise and subsequently difficult to manage). The aim of the two case studies is to learn how an innovative platform based on semantic Web technology can be useful in redesigning or modifying existing business processes.

3. Proposed solution
As a first example, the internal assistance process at Poste Italiane was examined. The objective is to improve efficiency and effectiveness in resolving issues and, where possible, reduce wait times. This is a highly structured process, where the current system handles large amounts of information. It is extremely important for a company such as Poste Italiane to preserve the functional continuity of services provided, which evidently results in a higher level of customer satisfaction and greater continuity in the development of the business.

There are essentially two channels through which notifications arrive: the Web and the call centre. It should be borne in mind that the amount of resources available to manage the entire process is limited when compared to the amount of existing work. The current process is structured on three main levels. Once the notification request has been received, a first level of analysis takes place. In this first phase of the total amount of requests received, 38% of problems are resolved at the first level. In case of a negative outcome, second level dispatching is effected, obtaining a resolution of the problem within 24 hours in 40% of cases. Any actions that require further technical checks find resolution within 48 hours in 5% of cases. If, however, the notification received cannot be resolved by Poste's internal staff, then it is sent to the maintainer for specialist on-site intervention.

The innovative solution proposed foresaw semantic technological intervention from the earliest notification phase. The value added that this type of technology can provide unquestionably lies in the automation and greater structure of managed information. The system should be able to analyse the problem from the first level and propose possible solutions. The platform is able to make semantic correlations between the information available in the databases. All this can be undertaken independently without having to instigate a notification. If the problems require specific action on the second or third level, then the system can guide the operator in the structured compilation of the notification starting from the signalling phase. Furthermore, the system, through semantic analysis, can also automatically allocate a first priority level. Another important aspect of using a platform such as this is undoubtedly the ease of integration at any level, since external maintainers can also very easily update the system without having to incur the onerous costs of integration with their own systems. Naturally, to create semantic correlations and structure semantic information, an appropriate ontology of the malfunction must be defined. In fact, a possible hardware issue is linked to the failure of delivering a particular service or type of service. Only by defining an adequate semantic tree will the system be able to make the appropriate correlations and support the employee in the formulation of the request for assistance. All this will certainly help the subsequent levels in interpreting and formulating a possible solution.

The other business process that the team worked on is the Intesa San Paolo start-up initiative. A future scenario where semantic technology may easily be applied, and can significantly contribute, is the evaluation of documents that make up the business plan.

Through examining the project in detail, the study group hypothesised the possible factors that enable a more efficient and focused analysis to achieve two objectives: first, to formulate the start-up to render the business plan more solid, constructing an efficient elevator pitch from the investor perspective; thereafter, presented as an environment for the best start-ups to encounter the best Italian and foreign investors.

Semantic Web offers a significant contribution, especially in the early stages, where the process of the initiative is structured, intervening in the following activities: scouting, gathering notifications, business plan analysis, training, deal line-up and arena. During the scouting of around 120 companies each year, semantic technology allows the banking group to assume an active position, not only in terms of receiving, but also in the search for start-ups in Italy and in new foreign markets. In the subsequent notification-gathering phase, the new technology offers support to the proposer in preparing the documents that constitute the business plan and a dedicated e-mail service based on the recognition of natural language. In the analysis phase, the Web search renders the entire information-gathering process on the entrepreneur's reputation, references, the industry and applied technology benchmarking more efficient. Training takes place in a classroom aimed at preparing business plan documentation and meeting investors. Thereafter, during the deal line-up, a panel of industry experts continues the technical feasibility study on the development and potential success of the venture. Finally, the best 10/12 companies undertaking real innovations are brought into the arena favouring meeting potential investors (including business angels, seed funds, corporate and venture capital).

The result of this activity is more thorough and active scouting and, consequently, a greater degree of detail of information on enterprises, while also reducing the cost and time of the search and selection processes.

In conclusion, providing contexts of higher uncertainty with greater operative capacity is a prerequisite for success. The amount of data is constantly increasing and a possible solution is the use of the semantic Web. This entails significant consequences such as changes in firm organisational forms and creating new positions and professionalism within business processes, both on ICT and functional levels.

The challenge of healthcare services: between process standardisation and service customisation

8

Sabina Nuti and Cinzia Panero

In many contexts, process standardisation and service customisation are considered rather opposed strategies in the search for competitive advantage: the former aimed at achieving economies of scale and experience to increase production efficiency and reduce costs, and the latter to achieve differentiation with respect to competitors based on "tailor-made" services to meet individual customer needs. In healthcare services, success depends on the ability to integrate these two strategies. No healthcare service is able to be effective unless it is targeted at the needs of individual patients. At the same time, the quality of services depends on the capabilities of professionals and organisations to provide healthcare pathways in line with international scientific evidence and to standardise processes according to clinical protocols defined by the scientific community. The service offer should hence be customised so that patients become the protagonists of their own healthcare pathway, but should also ensure that the best care (clinical appropriateness) is provided within the most appropriate setting for the best use of available resources (organisational appropriateness).

S. Nuti (✉)
Istituto di Management, Scuola Superiore Sant'Anna, Pisa, Italy
e-mail: s.nuti@sssup.it

C. Panero
Dipartimento di Economia, Università degli Studi di Genova, Genova, Italy
e-mail: c.panero@sssup.it

L. Cinquini, A. Di Minin, R. Varaldo (eds.), *New Business Models and Value Creation: A Service Science Perspective.* Sxi 8, DOI 10.1007/978-88-470-2838-8_8, © Springer-Verlag Italia 2013

8.1
The challenge of healthcare services between quality and financial sustainability

The strategies adopted by firms in the context of private sector manufacturing, but also in services, can be divided into two types.[1] Firms can either point towards cost leadership, through strategies aimed at achieving economies of scale and experience in order achieve market success in terms of efficiency (the best product at the lowest price), or focus on differentiation, trying to offer a product that meets individual customer needs through a process of customising their products and services (Porter 1985). Through these two alternative strategies, firms seek to secure a competitive advantage as well as endurance on the market.

Do these two alternative strategies also exist in the healthcare setting? Are the terms the same for healthcare firms? Are they the same for healthcare systems that include several firms with different roles, in some cases networked (Miolo Vitali and Nuti 2003), in others in competition with each other? The healthcare setting is in fact one of the few areas, together with education, where success depends on the capacity to integrate these two strategies, which are usually alternatives in other sectors.[2]

No healthcare service can be considered valid unless it is tailored to the needs of individual patients. The offer can hardly be standardised, because all users have their own characteristics and specificities, but at the same time, it must be based on standards to ensure the quality of the service provided.[3] In other sectors, increasing the quality of products often requires an increase in the resources used in the processes of change or in the production factors employed.[4] In healthcare, this is not always achieved, precisely because of the necessary combination of strategies.

Jarman's (2006) notable study on Medicare 2000 data analysed thousands of hospital admissions, comparing costs with the quality results expressed in terms of mortality in American hospitals (Fig. 8.1).

This comparison between mortality and cost, after an appropriate risk adjustment process, shows that there is no clear correlation between outcome and cost incurred.

This same analysis was carried out in Holland, reporting similar figures. In this reality, it was even calculated that 25% of national healthcare expenditure is due to "non-quality", namely, repeat admissions and re-hospitalisations due to complications, longer hospital stays due to bedsores, inappropriate hospitalisation for chronic

---

[1] Three, when considering the focus, nevertheless resulting from cost leadership or differentiation, but applied to a segment rather than the entire market.

[2] Indeed, Anderson et al. (1997) revealed that customisation and standardisation are two often-conflicting aspects of quality.

[3] This is the main difference compared to other sectors and services, where the redefinition of the service process, in a standardised way, is considered an essential element but only to increase productivity (Lovelock and Wirtz 2007).

[4] In fact, in an array of economic studies it is believed that the relationship between increasing customer satisfaction and productivity is negative: increasing satisfaction involves increasing the characteristics of the exchanged product and thus the costs (Griliches 1971; Lancaster 1979). There are also those (e.g., Fornell and Wernerfelt 1988) who observe that by increasing quality, and therefore customer satisfaction, operating costs associated with returns are reduced, or those who (Reichheld and Sasser 1990) emphasise greater loyalty and thus cost reductions resulting from future transactions and positive word of mouth.

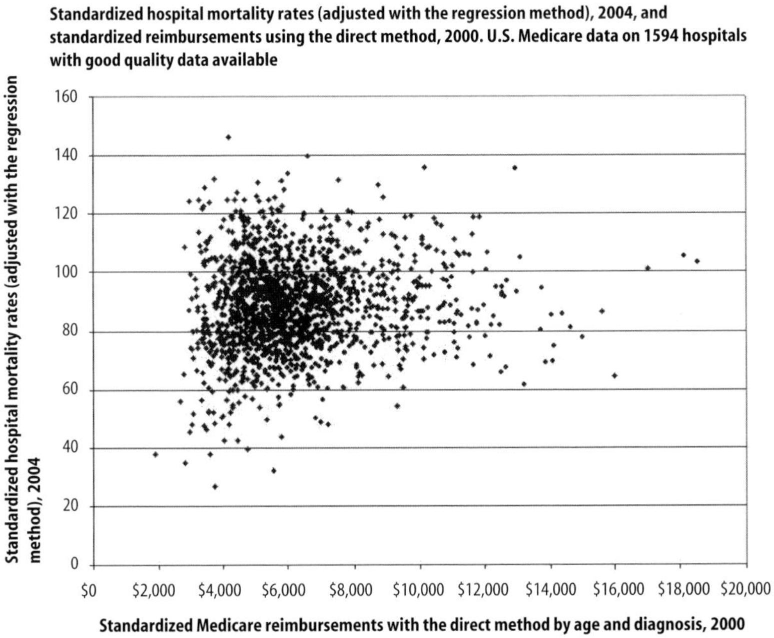

Standardized hospital mortality rates (adjusted with the regression method), 2004, and standardized reimbursements using the direct method, 2000. U.S. Medicare data on 1594 hospitals with good quality data available

y-axis: Standardized hospital mortality rates (adjusted with the regression method), 2004

160
140
120
100
80
60
40
20
0

x-axis: $0 $2,000 $4,000 $6,000 $8,000 $10,000 $12,000 $14,000 $16,000 $18,000 $20,000

Standardized Medicare reimbursements with the direct method by age and diagnosis, 2000

Fig. 8.1 Results of the American study on the relationship between reimbursements and mortality (Jarman 2006)

diseases that should be treated in other care settings and so forth (Berg et al. 2005). This shows that the improvement of quality in healthcare can frequently even result in cost containment. This hypothesis is supported by 2007, 2008 and 2009 data from the Tuscany region, where it clearly emerged – thanks to the adoption of a performance evaluation system (Nuti 2008) that monitors 130 performance indicators – that the local health authorities (LHA) with the best quality and efficiency results are also the most virtuous in terms of economic sustainability (Fig. 8.2).

The challenge of healthcare services starts from these premises and it is based upon the concept of clinical and organisational appropriateness (Nuti and Vainieri 2009). The term appropriateness refers to the capability to provide users with a tailored service taking into account their needs: "nothing more and nothing less" than is necessary to achieve the best results in terms of health. A service is appropriate precisely when everything that scientific evidence indicates as essential to obtain the best outcome is offered, but nothing more, because excess may even be harmful to health: hence, the best cure that can be offered to the patient (clinical appropriateness), in the most appropriate setting to ensure the best use of available resources (organisational appropriateness) (Hunter 1997; Brennan et al. 1991).

The service offer must be customised especially in terms of healthcare operator–user communication to ensure that patients become the protagonists of their own healthcare pathway, a determinant factor in the maximisation of health outcomes. Patients that are aware and involved in their healthcare pathway, who sense being

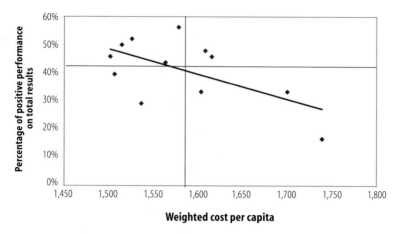

Fig. 8.2 Relationship between cost per capita and percentage of positive performance results on total results of the performance evaluation system in Tuscany

taken care of by the medical staff, are better able to follow drug prescriptions and clinical trials and have a greater probability of recovering a better state of health (Stewart 1989; Guldvog 1999). At the same time, the proposed healthcare pathways must follow clinical protocols and thus be "standardised" in order to obtain the best results in terms of patient health, in line with international evidence.

The next section develops the argument as follows. Starting from the definition of the offer, which must necessarily be based on user needs, Sect. 8.2 describes the characteristics and the role this plays within healthcare services. Section 8.3 discusses to what extent variability currently present in the volume and mix of performance actually responds to different user needs, the problems to be overcome and the management methods to use (Sect. 8.4) with recourse to the Tuscan oncological pathway as an example of the combination of the two strategies outlined above (Sect. 8.5).

8.2
Characteristics of healthcare services and user roles

The World Health Organization, which defines healthcare services as "all services related to the diagnosis and treatment of disease, or the promotion, preservation and restoration of health" in 2000 challenged healthcare systems to ensure patient responsiveness, namely, care that is not strictly health related (respect for persons, confidentiality, choice of service provider).

This challenge appears complex, considering the particularities that characterise the healthcare services sector with reference to demand, supply and services exchanged (Arrow 1963), which, by distancing it from perfect competition, render supply and use difficult. With reference to demand, and hence the user of healthcare services, the most critical aspects are largely informative. Thus, as with all services, the patient cannot know the quality of the healthcare service[5] in advance since it is intangible and simultaneously produced and consumed. Arrow (1963) shows that information represents a fundamental problem in the healthcare market, noting that "the uncertainty concerning the quality of the product is perhaps more intense in this case than in any other".

The information issue involves a number of aspects in the patients' selection process. First, in the initial phase of defining their needs, since they are often unable to recognise the presence of symptoms and disease, and of deciding on whether to consult a doctor. Alternatively, the information problem can arise with the choice of healthcare facility and treatments to undergo once aware of a state of illness, and finally to the level of the decision on whether or not to comply with the recommended behaviour. Illness, especially if serious, is furthermore an exceptional episode of human existence where the individual's life can also be at stake: it is therefore difficult for the consumer to make rational decisions. It is also worth noting that whilst in most consumption processes people can learn from their own experience or from that of others (Arrow 1963), in this case, this may not be possible and hence the lack of previous experience is added to the uncertainty of the outcome.

The uncertainty that characterises healthcare services on the side of the patient and that of the doctor is very different for both subjects involved in the transaction (Arrow 1963), giving rise to information asymmetry (Miolo Vitali and Nuti 2004; Dirindin and Vineis 1999). Because scientific knowledge is complex, the information held by the doctor is far superior to that of the patient and both parties are aware of this. This particular situation affects the doctor–patient relationship, often resulting in the latter's strong sense of dependence (Miolo Vitali and Nuti 2004), since healthcare services are often provided to people in circumstances of unease. Information asymmetry moves the relation onto the trust component, which increases with the reduction of the evaluation component, thus replacing information. The healthcare market is therefore characterised by actors who make decisions based on incomplete, uncertain and asymmetric information (Dirindin and Vineis 1999). This asymmetry, unless corrective action is taken, is likely to increase with the scientific and technological progress of medicine (Brenna 1999).

The difficulties of healthcare service users are exacerbated by the characteristics typical of the healthcare service offer and the specific nature of the object of exchange. With reference to the offer, it has been observed that while a market, in order to be competitive, requires the presence of multiple manufacturers, the health sector is characterised both by the limited number of supply structures, due to entry barriers to the profession such as professional qualifications and specialisation, aimed at ensuring quality (Brenna 1999), and the cost of education, often with restricted access (Arrow 1963; Dirindin and Vineis 1999). Lack of competition is also due to economies of scale in the health sector resulting from the need for infras-

---

[5] It is worth noting, in particular, that health services are similar to experience goods (Nelson 1970), i.e., goods whose quality can be known through actual consumption or, often, credence goods (Darby and Karni 1973), impossible to judge even after prolonged use.

tructures and instruments that are characterised by significant indivisibility and thus increasing returns to scale, albeit only up to a certain level (Drindin and Vineis 1999).

These characteristics of the offer are accompanied by some specificities relating to the good exchanged, namely:

1) The good exchanged is a service and is, as such, marked by well known characteristics of intangibility, inseparability, perishability and heterogeneity (Shostack 1977; Grönroos 1978; Parasuraman et al. 1985; Stanton and Varaldo 1986; Cozzi and Ferrero 1996; Lovelock and Wirtz 2007) (see Chap. 6).

2) Healthcare services, in particular, are extremely heterogeneous; indeed, since their demand is derived – inasmuch as they are not requested per se but considered necessary for positive effects on health (Brenna 1999) – they must be customised to be effective (Dirindin and Vineis 1999). In fact, every individual has different needs depending on the disease and the complications that arise.

3) Externalities, or effects on third parties in terms of costs or benefits resulting from production or consumption of a good without any payment in money terms. Healthcare services generate many, especially positive, externalities (i.e., they produce increases in marginal utility also for individuals other than the consumer of the service, as in the case of vaccinations[6]) (Culyer 1971), but also negative externalities (such as, for example, hospital infections, or noise pollution in neighbourhoods that are adjacent to emergency rooms). The price system in these cases is not able to charge or accredit to the actors the external costs and benefits they produce. Therefore, recourse is often made to public intervention, aiming to stimulate the production of positive externalities (subsidising activities that generate them) and to discourage negative externalities (regulating them) (Dirindin and Vineis 1999).

Externalities and the existence of monopolistic positions constitute, along with characteristics relating to the demand described above (imperfect knowledge), a source of market failure and affect the way in which the supply of the service itself is organised, thus justifying public intervention.[7] Such intervention is motivated not only by reasons of efficiency linked to potential market failure, but also by considerations relating to the information asymmetry that characterises the patient–healthcare relation and the need to ensure fairness, i.e., overcoming the constraints that impede access to healthcare for the most fragile, such as the have-nots or people with a low level of education. Motivation is necessary to activate perception and action (Rosenstock 2005); people, therefore, who are not particularly concerned about their health will probably not receive any information that impacts on it and, if perceived, would not be able to learn, accept or use it.

---

[6] Other positive externalities are the dissemination of scientific knowledge in the medical field, the discovery of a new diagnostic technique and the identification of risk factors. While the party carrying out the research or training activity sustains the costs related to project implementation, society takes advantage of the external benefits.

[7] It should be noted that health services are public goods (i.e., non-rival and non-excludable), but goods worthy of special protection because they are socially meritorious (Dirindin and Vineis 1999; Brenna 1999).

Indeed, many studies on healthcare services over time have found significant social inequalities. In terms of mortality, if the link between income and health is controversial (Dirindin and Vineis 1999; Nuti and Barsanti 2010), only initially highlighting a negative relation,[8] then between education and mortality it is instead broadly confirmed. Back in the 1970s, Valkonen (1992) showed that mortality was reduced for each additional year of education. This result is understandable considering that a high level of education encourages greater attention to risk factors and symptoms and, in the event of illness, greater ease of access to different treatment options; generally, therefore, a higher level of health ensues.

Education also has effects on the health of citizens, as confirmed by the ISTAT multipurpose survey (ISTAT 2007), which, using education as an indicator, demonstrates the presence of strong social inequalities. People with low educational qualifications are always those with the worst health conditions, both in terms of perceived health and chronicity.[9]

Concerning the use of services, Rosenstock (2005) previously demonstrated that some generalisations can be made about the association between personal characteristics and the use of preventive and diagnostic services: these are mainly used by the young or middle-aged who are relatively more educated and have a higher income level. These results are also confirmed by more recent research in relation to "training" (such as attending a course of antenatal classes) and hospitalisation itself. From a survey conducted in Tuscany in 2005 and repeated in 2007 (Nuti and Barsanti 2006; Nuti et al. 2009) on a sample of women who had given birth in the previous months, it emerged that antenatal classes, considered a useful tool to increase the knowledge of mothers, were attended by 60% of primiparous women, but 70% of these were almost exclusively graduates, with a total absence of those with only an elementary school leaving certificate or without any qualifications, i.e., the most vulnerable.[10] Also in Tuscany, longer hospitalisation by education level standardised by age (Barsanti 2010a, 2010b), emergency instead of planned hospitalisation (indicative of the presence of disease or more serious difficulties to access appropriate care) and hospitalisation for chronic diseases (heart failure, diabetes, COPD, pneumonia) are more common in less educated populations, albeit varying greatly among different health authorities (Barsanti 2009; Barsanti et al. 2009).

Finally, studies on satisfaction arising from the use of healthcare services – an important aspect since satisfied patients are more inclined to comply with prescriptions (Guldvog 1999) and to take an active role in their healthcare process (Donabedian 1988) – highlighted the importance of education level, but with an inverse relationship: the most educated present the highest levels of dissatisfaction (Hall and Dornan 1990). These results are also confirmed by subsequent studies (Panero et al. 2010),

---

[8] More specifically, the relation in turn is negative in underdeveloped countries, thus as income increases mortality decreases significantly, while in developed countries, life expectancy increases with the decline in inequality in the community rather than with the increase of wealth.

[9] The ISTAT survey shows that those who have at most an elementary school leaving certificate and declare they are sick, or affected by chronicity, are in fact up to three times more numerous than graduates.

[10] A woman in labour with a low educational level is in fact more prone to social problems and difficulties in accessing services, entailing the greater probability for a child to incur health risks and food shortages.

relating to general medical services: a high level of education shows a negative relationship with respect to service satisfaction (consistent with the higher expectations of this type of patient). This study also highlights the difficulty patients have in assessing the service, as testified by the significance that some organisational aspects assume, in addition to the doctor–patient relationship (Murante et al. 2010), such as the doctor's use of medical health records and, in a negative sense, waiting time or the absence of continuity of care in case of specialist needs, elements on which patients can more easily express their opinions since they are easier to preside over (Miolo Vitali and Nuti 2004).

We can therefore conclude that, considering the specificity of healthcare services and the user's role, the provision of healthcare services must respond to two types of responsibilities (Nuti 2008; Nuti and Vainieri 2009): the organisational, concerning the use of scarce resources in a sector where the market may not be able to provide appropriate responses, and the clinical, concerning the effectiveness of treatments, taking into account the condition of information asymmetry that characterises the doctor–patient relationship (Nuti and Barsanti 2006; Mengoni et al. 2010).

8.3
Service characteristics and performance variability

In the previous section we stressed the importance of considering the patient's role in the definition of an adequate provision of healthcare services and we now look in more detail at how the variability that occurs today in the volume and mix of services provided needs to effectively respond to the differentiated needs of users. An analysis of the data available at national level (Ministero della Salute 2010) suggests considerable variability between the results obtained by the different regional health systems and among providers within each region. Take, for example, a regional comparison of the capability of providers to intervene surgically within two days for hip fracture patients. International scientific evidence points to the need to intervene promptly in order to facilitate the next phase of recovery but also to reduce the risk of mortality. In 2008 (Nuti 2010), the situation was rather variegated on the national scene, undermining the principle of access equality, which should be the very foundation of our healthcare system. The problem of variability appears to be a critical point not only in the north–south comparison (Fig. 8.3). Actually there is also variability among best-practice regions (Fig. 8.4).

Is this variability plausible and acceptable in our country? If the key concept of quality in healthcare is "appropriateness", which is to ensure patients nothing more and nothing less than that which is required to effectively respond to their specific needs, the issue of variability could be the result not of a process driven by customising the service but an indication of the lack of provision in some cases or an excessive, and sometimes harmful, response in others. Excessive variability thus becomes an indication of non-equity and causality in the delivery process.

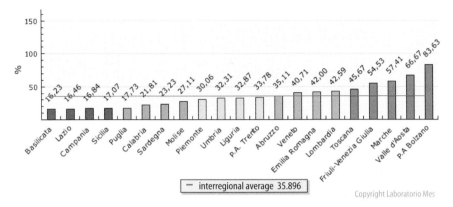

Fig. 8.3 Percentage of femur fractures operated on within 2 days in Italian regions

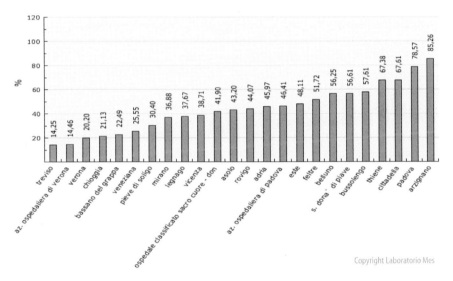

Fig. 8.4 Percentage of femur fractures operated on within 2 days in the Veneto Region (Laboratorio Mes)

   This frequently occurs because the necessary sharing of clinical protocols to be adopted among health professionals is lacking, as is the capacity for teamwork, with an approach that goes beyond the offer of the single service in favour of a set of services that structurally and organisationally make up the complete service provision process. Lee (2010) explains the difficulty that medical professionals have in collaborating and sharing appropriate patient responses with their propensity for professional autonomy, a crucial principle to pursue service quality. Actually, professional autonomy is not synonymous with healthcare quality. Appropriate responses to patient needs increasingly depend on a team of professionals and their capability to think in terms of processes instead of single service provision.

8.4
Process standardisation and service customisation: healthcare pathways

Since the early 1990s, several authors have developed process management studies, defining the process itself as a "set of logically related tasks performed to achieve a defined outcome" (Davenport and Short 1990), or "a set of activities using one or more types of input creating an output that has value to the customer" (Hammer and Champy 1993), and finally, as an entity capable of "capturing the inter-functional interdependencies and linking improvement efforts to strategic objectives" (Kaplan and Murdock 1991).

In functional-type organisations the concept of "workflow" or flow of activities already existed. Consider, for example, the assembly line. So what are the differences with respect to process management? Here are the four key differences:

1) In the first place, the focus is on the customer: everything has meaning and value to the extent that it directly or indirectly contributes to determining the satisfaction of end customer needs.
2) The focus on organisational effectiveness over organisational efficiency. Cost containment and price minimisation are no longer the only winning factors but the quality and customisation of goods and services provided are equally important.
3) Products and services are placed on the market through activity flows that crossover organisational units. Hierarchy as a coordination mechanism between organisational units is not sufficient to ensure the flow of information from the customer to all the functions involved in the creation of value for customers. Firms, therefore, adopt mechanisms and tools to facilitate cross-communication between the direct functions to ensure the speed and timeliness of responding to customers.
4) Finally, the need to preside over the determinants of long-term business success based on the capacity to respond to customer needs in an innovative and flexible way.

To initiate process management within the organisation, the starting point is the analysis of the characteristics and lifestyles of customers/end users: understanding who they are, and what their needs and their wants are. This is a prerequisite for formulating the offer in a way that is personalised and fully responsive to the demand and, if possible, even in a "proactive" way, i.e., anticipating the needs themselves. The capability to analyse, attend and listen to user needs, both explicit and implicit, becomes the firm's distinctive competency, because this allows definition of the elements of the service system that increase value for the customer. On this added and differential value, firms can build both trust and continuity with the customer as well as a competitive advantage.

From the comparison between customer/user needs and the firm's offer (Casati 1999; Miolo Vitali and Nuti 2004), the strengths and weaknesses, or aspects of excellence and gaps in the offer, are identified. The phases of analysis and process mapping are formulated on this basis (Merli and Biroli 1996; Candiotto 2003), with the objective of recognising on one hand the "determinants" of value created for the user, i.e., the critical activities/processes, and on the other, identifying those activities that do not create value. These latter, in turn, may either be useful for internal organisation and should therefore be maintained, or are completely useless and can thus be eliminated. With this in mind, firms can identify targets for improvements and develop strategic quality plans (Hammer and Champy 1993; Pierantozzi 1998).

In healthcare, management processes assume specific characteristics that require due attention: What are the dimensions of a business process in terms of health? The answer may differ depending on the perspective chosen. For the past few years, even in Italy, thanks to the dissemination of evidence-based medicine, guidelines and treatment protocols have been introduced for many diseases that facilitate sharing treatment methods and care pathways among doctors. This initial result, while important, does not imply working in terms of "processes". In fact, from the therapeutic clinical path, the service from the patients' perspective, and not only their care, must be developed, thus entailing thinking in terms of the "healthcare pathway".

According to Article L.R.22/00. §2 letter m (taken from art. 4 of 1.40/2005) (L.R. 2005), a healthcare pathway is defined as "the result of an organizational method that promptly ensures to citizens, in a coordinated, integrated and programmed way, informed access and appropriate and shared use of local healthcare and hospital network services". This is therefore the path that enables citizens to find a response to specific health problems. The goal for users is to acquire "value", step-by-step and activity-by-activity, in terms of quality. In other words, responsiveness to their health problems. To rebuild the healthcare pathway requires starting from users, reviewing the entire service delivery process "through the patients eyes"[11] by way of their experience.

It is obvious that in many cases users find themselves in conditions of asymmetric information and are unable to clearly identify what is good for them from a clinical health standpoint since they do not have the necessary skills and only medical staff hold the knowledge to guide the treatment. But the point is not to replace doctors in their prerogatives as much as to enhance their work with the organisation of a pathway, taking into account the specific characteristics of patients, valorising the activities carried out by medical staff for the benefit of users. In business terms, we could refer to a Healthcare Value Chain (Burns et al. 2002), namely, where the ultimate aim is the improvement of the user's wellbeing. This is therefore about finding the ways in which to insert the patient's experience into the delivery process, surpassing a Taylorist logic of work organisation in favour of a structure that takes into account the centrality of the user. In the realty of our healthcare structure today, talking about a care pathway is not feasible since it is users themselves, not the healthcare organisation, that liaise between the different components and phases of the service. The challenge is instead to review the delivery mode, bearing in mind,

---

[11] We recall in this respect, the health management text of Gerteis et al. (1993), widely diffused in the United States.

stage by stage, the patient's needs and propose a path where the coordination of the offer and the continuity of care are the objectives that the healthcare facility itself presides over.

8.5
Comparison of patient needs with the provision of healthcare services: the case of the oncological pathway in Tuscany

During the 2000s, the Tuscany Region adopted a series of measures to ensure cancer patients a service capable of responding to the complex and highly emotional care needs related to cancer treatment. In particular, it was determined that the service should be articulated and coordinated into a patient-centred "care pathway" (thus ensuring patient's care), characterised by the adoption of shared clinical protocols based on scientific evidence and professional integration and continuity of care between health authorities, hospitals and the territory in a "network" logic.[12],[13]

To verify compliance with the above-indicated objectives, the Region of Tuscany, through the Tuscan Cancer Institute in collaboration with the Laboratorio Management e Sanità of the Scuola Superiore Sant'Anna, called for an assessment of the quality of its services by monitoring whether the healthcare pathways designed by local health authorities (LHA) observed the Tuscany Cancer Institute's clinical protocols outlined on the basis of international scientific evidence and therefore "standardised", as well as the perception of the care received from listening to patients and their needs and expectations, on the basis of which to reformulate the care offer. This result was achieved first through a quantitative-type telephone survey (Nuti and Murante 2008) and thereafter a qualitative survey based on the analysis of evidence that emerged from focus groups with cancer patients (Nuti et al. 2010).

This second monitoring and mapping phase was accompanied by the ideal pathways (i.e., as defined by the clinical recommendations of the Tuscan Cancer Institute) and the actual pathways (as currently offered by LHA), thus exposing the gaps between them. Gaps were classified as:

1) Process gaps (PX in Fig. 8.5), arising from the absence in the actual pathway of some phases or organisational methods foreseen by the ideal diagnostic-therapeutic path, or by their lack of perception by patients.
2) Timeline gaps (TX Fig. 8.5), namely, higher than expected delays as foreseen by the ideal path or treatment needs.

---

[12] The Tuscany Cancer Network was established by the Regional Council Decree No 532 of May 27, 2002.

[13] The clinical protocols were made available to all local health professionals and through publication of the "Clinical Recommendations for Main Solid Tumours" (Tuscan Cancer Institute 2005), elaborated by over 400 oncological healthcare workers.

3) Coordination and relational gaps (Rx Fig. 8.5), related to the lack of continuity of the pathway, and the lack of organisation or communication errors between the different actors involved in the process, including patients.

The same classification was used to expose the criticalities described by patients in focus groups: it was thus also possible to synthesise the deviations of the pathway organised by the heath authorities with respect to user needs.

As can be seen from the summary of the gaps at regional level (Fig. 8.5), it is clear that, despite publication and sharing of recommendations and clinical protocols, which constitute an important step in the dissemination of the most appropriate therapies, there is still a great deal to be done to ensure that patients are guaranteed a consistent offer throughout the territory that is appropriate to their needs: gaps in processes, such as timing and coordination between facilities and professionals, are in fact still very numerous. However, the methodology adopted formalises both ideal pathways and those actually offered by health authorities by classifying the possible criticalities in a concise but understandable way, offering the professionals involved in the healthcare pathway the opportunity to reflect and thus enable them to identify how to intervene and improve the overall service.

The hypothesis of adopting methodologies of this type, and replicating them over time, would seem particularly useful: these are important instruments to perceive the systematic quality of pathways organised by LHA. In this way, the degree of improvement achieved in the organisation of pathways can be verified in terms of both clinical standardisation and customisation, further verifying through the analysis of the experience of patients whether and to what extent the organisational solutions implemented thus far have been observed and appreciated by patients in their experiences and have not remained unheeded despite best intentions.

8.6
Conclusions

The healthcare environment is characterised by a highly professional and, at the same time, a highly complex context. In organisational terms, this is a very challenging and constantly evolving environment. Technological innovation and scientific research on the one hand, and the natural growth of the population's health needs on the other, require a continuous review of healthcare service provision to identify viable solutions in therapeutic terms that are sustainable in financial terms.

The combined study of user demand, their habits, socio-economic conditions and the method to introduce the optimisation of management processes and the rationalisation of activities in line with scientific evidence is an ongoing challenge for healthcare management. In this sense, the study also highlights how, in a context such as healthcare, marketing methods can be useful tools that allow guidance of the offer towards patient needs in order to avoid wasting resources. The study of user behaviour and needs enables identification of the elements to be considered when

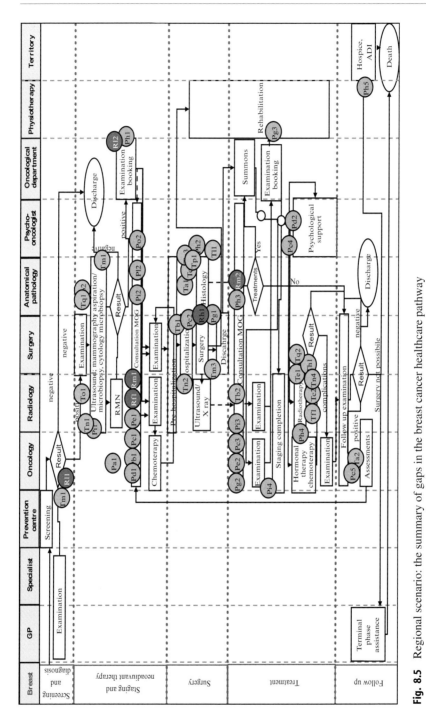

Fig. 8.5 Regional scenario: the summary of gaps in the breast cancer healthcare pathway

Notes: $P_x 1 \ldots P_x n$ indicate the process gaps of hospital x from 1 to n; $T_x 1 \ldots T_x n$ indicate the timeline gaps of hospital x from 1 to n; $R_x 1 \ldots R_x n$ indicate the relation and coordination gaps of hospital x from 1 to n. The hospitals (x) are indicated with lowercase letters defining appropriate healthcare service provisions (see also Chap. 9). This is especially important in communication processes between patients and health professionals so that the latter assume a proactive role, especially towards frailer patients, in providing prevention, care and assistance. Personalisation of the service must manifest itself in the capacity to adapt the service to the specificities of users, yet not give rise to unjustifiable variability in the delivery process validated by clinical protocols and the guidelines proposed by scientific societies and international evidence.

The success and excellence of healthcare organisation services rests on the combination of these two strategies.

References

Anderson EW, Fornell C, Rust RT (1997) Customer satisfaction, productivity and profitability: differences between goods and services. Market Sci 16:129–145

Arrow KJ (1963) Uncertainty and the welfare economics of medical care. Am Econ Rev 53:941–973

Barsanti S (2009) La specificità degli strumenti per misurare e valutare l'equità. In: Nuti S, Vainieri M (eds) Fiducia dei cittadini e valutazione della performance nella sanità italiana. Edizioni ETS, Pisa

Barsanti S (2010a) La valutazione della capacità di perseguire le strategie regionali. In: Nuti S (ed.) Il sistema di valutazione della performance della sanità toscana Report 2009. Edizioni ETS, Pisa

Barsanti S (2010b) Salute, equità e sviluppo economico. In: Frey M, Meneguzzo M, Fiorani G (eds) La sanità come volano dello sviluppo. Edizioni ETS, Pisa

Barsanti S, Tedeschi P, Nuti S (2009) Cronicità e disuguaglianze in salute: spunti per riconfigurare l'assistenza in base alle determinanti sociosanitarie presenti in Regione Toscana. VII Congresso nazionale CARD, 19–21 March, Pisa

Berg M, Meijerink Y, Gras M et al (2005) Feasibility first: developing public performance indicators on patient safety and clinical effectiveness for Dutch hospitals. Health Policy 75:59–73

Brenna A (1999) Manuale di economia sanitaria. CIS Editore, Milan

Brennan TA, Leape LL, Laird NM et al. (1991) Incidence of adverse events and negligence in hospitalized patients: results of the Harvard Medical Practice Study I. N Engl J Med 324:370–377

Burns LR, DeGraaf RA, Danzon PM et al. (2002) The Wharton School study of the health care value chain. John Wiley & Sons, New York

Candiotto R (2003) L'approccio per processi e i sistemi di gestione per la qualità. Giuffrè Editore, Milan

Casati G (ed.) (1999) Il percorso del paziente. La gestione dei processi in sanità. Egea, Milan

Cozzi G, Ferrero G (1996) Marketing. Principi e tendenze evolutive. Giappichelli Editore, Turin

Culyer AJ (1971) The nature of the commodity 'health care' and its efficient allocation. Oxford Econ Pap 23:189–211

Darby MR, Karni E (1973) Free competition and the optimal amount of fraud. J Law Econ 16:67–88

Davenport TH, Short JE (1990) The new industrial engineering: information technology and business process redesign. Sloan Manage Rev Summer:11–27

Regional Council Decree No 532 of May 27, 2002. Rete oncologica regionale: Prime deter-
minazioni in applicazione del P.S.R. 2002–2004.

Dirindin N, Vineis P (1999) Elementi di economia sanitaria. Il Mulino, Bologna

Donabedian A (1988) The quality of care. How can be assessed? JAMA 260:1743–1748

Fornell C, Wernerfelt B (1988) A model for customer complaint management. Market Sci
7:271–286

Gerteis M, Edgman-Levitan S, Daley J, Delbanco TL (1993) Through the patient's eyes: understanding and promoting patient-centered care. Jossey-Bass, San Francisco, CA

Griliches Z (1971) Prices indices and quality change. Harvard University Press, Cambridge,
MA

Grönroos C (1978) A service-oriented approach to marketing of services. Eur J Mark 12:588–
601

Guldvog B (1999) Can patient satisfaction improve health among patients with angina pec-
toris? Int J Qual Health C 11:233–240

Hall JA, Dornan MC (1990) Patient sociodemographic characteristics as predictors of satisfaction with medical care: a meta analysis. Soc Sci Med 30:233–240

Hammer M, Champy J (1993) Reengineering the corporation: a manifesto for business revolution. Harper Collins, New York

Hunter DJW (1997) Measuring the appropriateness of hospital use. Can Med Assoc J 157:901–
902

ISTAT (2007) Condizioni di salute, fattori di rischio e ricorso ai servizi sanitari. ISTAT, Rome

Istituto Toscano Tumori (2005) Raccomandazioni cliniche per i principali tumori solidi

Jarman B (2006) Using health information technology to measure and improve healthcare
quality and safety. Paper presented at the 23rd Annual International Conference on the
International Society for Quality in Health Care, London

Kaplan RB, Murdock L (1991) Core process redesign. McKinsey Q 2:27–43

Lancaster K (1979) Variety, equity, efficiency. Columbia University Press, New York

Lee TH (2010) Turning doctors into leaders. Harvard Bus Rev April:50–58

Lovelock C, Wirtz J (2007) Marketing dei servizi. Risorse umane, tecnologie, strategie. Pear-
son, Prentice Hall

L.R. (22/2000) Riordino delle norme per l'organizzazione del servizio sanitario. 8 March,
n. 22

L.R. (2005) Disciplina del servizio sanitario regionale. 24 February, n. 40

Mengoni A, Murante AM, Nuti S, Tedeschi P (2010) Segmentazione e marketing per la sanità
pubblica. Mercati e competitività 1:119–138

Merli G, Biroli M (1996) Organizzazione e gestione per processi. ISEDI, Turin

Ministero della Salute (2010) Progetto SiVeAS. http://www.salute.gov.it

Miolo Vitali P, Nuti S (eds) (2003) Ospedale in rete e reti di ospedali: modelli ed esperienze
a confronto. Franco Angeli Editore, Milan

Miolo Vitali P, Nuti S (eds) (2004) Sperimentazione dell'Activity Based Management nella
sanità pubblica: l'esperienza dell'Azienda USL 3 di Pistoia. Franco Angeli Editore, Milan

Murante AM, Panero C, Nuti S (2010) L'esperienza dei cittadini del servizio di medicina
generale: come la comunicazione influenza la relazione medico–paziente. Quattro regioni
a confronto. VIII Congresso Nazionale CARD, 16–18 September, Padua

Nelson P (1970) Information and consumer behaviour. J Polit Econ 78:311–329

Nuti S (ed.) (2008) La valutazione della performance in sanità. Il Mulino, Bologna

Nuti S (ed.) (2010) Il sistema di valutazione della performance dei sistemi sanitari regionali.
Primi indicatori ministeriali. Anno 2008. Ministero della Salute. 21 April, http://www.
salute.gov.it/dettaglio/phPrimoPianoNew.jsp?id=273&area=ministero&colore=2

Nuti S, Barsanti S (2006) L'accesso al percorso materno infantile. Salute e Territorio 158:303–305

Nuti S, Barsanti S (2010) Cronicità e spesa sanitaria. In: Frey M, Meneguzzo M, Fiorani G (eds) La sanità come volano dello sviluppo. Edizioni ETS, Pisa

Nuti S, Murante AM (2008) L'esperienza e la soddisfazione dei pazienti oncologici per i servizi sanitari ricevuti in Toscana. In: AA.VV. (eds) La valutazione della qualità nella rete oncologica toscana. Dalle raccomandazioni cliniche ITT agli indicatori di percorso assistenziale. Giunti Editore, Florence

Nuti S, Vainieri M (eds) (2009) Fiducia dei cittadini e valutazione della performance nella sanità italiana. Edizioni ETS, Pisa

Nuti S, Bonini A, Murante AM, Vainieri M (2009) Performance assessment in the maternity pathway in Tuscany Region. Health Serv Manage Res 22:115–121

Nuti S, Calabrese C, Panero C (eds) (2010) Confronto tra bisogni del paziente e offerta sanitaria per il miglioramento organizzativo del percorso oncologico. Economia Sanitaria, Milano

Panero C, Murante AM, Perucca G (2010) The patient needs and the answer of general practitioner: the Italian citizens experience. In: Testi A et al. (eds) Operations research for patient-centered health care delivery. Proceedings of the XXXVI International ORAHS Conference. Franco Angeli Editore, Milan

Parasuraman A, Zeithaml VA, Berry LL (1985) A conceptual model of service quality and its implications for future research. J Mark 49:41–50

Pierantozzi D (1998) La gestione dei processi nell'ottica del valore: miglioramento graduale e reengineering: criteri, metodi ed esperienze. Egea, Milan

Porter ME (1985) Competitive advantage: creating and sustaining superior performance. The Free Press, New York

Reichheld FF, Sasser WE (1990) Zero defections: quality comes to services. Harvard Bus Rev 68:105–111

Rosenstock IM (2005) Why people use health services. Milbank Q 83:1–32. Reprinted from Rosenstock IM (1966) The Milbank Mem Fund Q 44:94–124

Shostack GL (1977) Breaking free from product marketing. J Mark 41:73–80

Stanton WJ, Varaldo R (1986) Marketing. Il Mulino, Bologna

Stewart M (1989) Which facets of communication have strong effects on outcome: a meta-analysis. In: Stewart M, Roter D (eds) Communicating with medical patients. Sage, Newbury Park, CA

Valkonen T (1992) Socioeconomic differences in mortality 1981–1990. Central Statistical Office of Finland, Population, 8, Helsinki

INNOVATION LAB
Engineering new products for secure multimedia telecommunications networks

MAINS Master, academic year 2009/2010
People and companies involved in the InnoLab:
Students: Caterina Cinquini, Giulio Giovannetti and Samuela Locci
Companies: Ansaldo Energia, Elsag-Datamat, Finmeccanica Group Services and SIA-SSB
Professors: Alberto Di Minin and Andrea Piccaluga

1. The problem
The team was tasked with the development of a new product engineering process for secure multimedia telecommunications networks. The process would bring innovative elements with respect to the solutions proposed by standard R&D-engineering-pProduction as well as their potential viability in the firm. The primary objective was to support a generic company in entering a highly competitive market such as products for telecommunications networks. The overall objective was declinated into a more operative objective to be achieved in the reduced operating time available.

The team set out to design an engineering process of new standard products that would ensure high performance and allow a high level of customisation. The project was to eventually integrate an organisation that supports the process, a monitoring system that guarantees results and information systems that facilitate, through the standardisation of communications, activities between the actors and the process functions.

2. Work methodology
In the first instance, the team analysed the relevant literature on the development of new product processes and consumer telecommunications equipment (such as decoders for television applications, or switches, bridges and routers for networking applications).

The analysis, with respect to the main players in the market, highlighted two key drivers, namely the security provided by products and their ability to support multimedia content, which are envisaged as fundamental to competitiveness within the market of reference.

These characteristics are evolving very quickly following the ever-changing application standards that from time to time become available on the market. Effectively implementing these characteristics greatly complicates the products and, above all, requires that the engineering process ensure very short time to market.

In particular, overall trends were analyzed in terms of volume and turnover in the target market, the presence of leading players and their market shares by region as well as the strategies implemented by those market players who play the role of follower in order to identify a positioning in line with the firm's internal know-how and the strategic choices aimed at ensuring production efficiency.

Defining the company's strategic positioning with respect to the implementation of secure multimedia products for telecommunications networks was a fundamental aspect because it allowed the team to define the ensuing strengths that the process must achieve in order to sustain the business strategy.

In particular, the analysis carried out on the telecommunications product market identified two main competitive drivers in relation to which the context and players of reference were segmented: in turn, the level of customisation and overall performance of the products were evaluated in terms of safety, quality and integration.

The pathway to undertake, complete and enable an innovative process is often long and difficult to implement. In addition, to redefine a business process, the effort and investments required are often very high and the results frequently do not justify them. Undertaking a process innovation project requires maximising the likelihood that the intervention is completed with the expected results, both on schedule and within budget.

3. Proposed solution

The solution that the team arrived at entails a process that is defined in all activity phases and the timelines that govern it. The activities and phases were clearly defined in terms of responsibility allocated to individual business functions while identifying all the roles essential to the execution of the process. For each of these roles, the technical and managerial characteristics were delineated that would ensure performance of duties to the best ability. Thereafter, the most appropriate levers to be assigned to the actors of the process to maximise their chances of success were outlined in greater detail.

Once the activities were mapped and the job descriptions defined, the team proceeded with designing a business organisation to house the newly identified professionals and provide them with the desired levers.

At this point, the process had taken shape and the fine-tuning stage was able to commence. During this stage, the process was simulated a number of times, in situations of increasing detail, periodically highlighting possible improvements.

Once the desired level of quality had been achieved, the team proceeded with the definition of support tools for the execution of the process comprising a monitoring system and an information system.

As concerns the monitoring system, first a set of Key Performance Indicators (KPIs) were designed that could monitor the critical dimensions of

this process. The KPIs were then assigned to each of the previously identified roles, with the dual purpose of using them as a criterion for evaluating
own performance and to provide a useful tool for the rapid identification of
abnormalities.

Finally, concerning the information system, the team restricted itself to
defining the functional characteristics that would have to adequately support
all the actors of the process and to compensate for the deficiencies presented
in a manual execution of the process itself.

The new secure telecommunications and networking products engi-
neered according to the new model architecture, governed by the introduction of process owners as key figures, precise rules governing the organisation as well as careful research of performance parameters with the support of a customised information system, provided the real innovation to the
standard models.

Home healthcare services: an educative case for the development of a "service-dominant logic" approach in the marketing of high-tech services

9

Giuseppe Turchetti and Elie Geisler

Home healthcare services rely on clinical and administrative technologies generally known as "telemedicine". The purpose of this chapter is to describe a case in the implementation of home care initiated by a major urban trauma hospital for patients with chronic diseases, thus to make the case for the use of service-dominant logic in the implementation of technologies in the service sector. The findings from this study are twofold. First, we found that the main barriers to the implementation of technology-based home healthcare services are not technological but anchored in the logic of marketing services to patients. Secondly, we concluded that by employing service-dominant logic in the provision of technology-based home-care services we could increase the pace of implementation of home care and the value of the services provided.

The relevance of the case study presented in the chapter is that of a lesson learned from the health-care services sector, but applicable to other sectors of the service economy.

9.1
Context and objectives

Healthcare is one of the top issues that all industrialised countries have to deal with today, and more and more will have to address in the coming years. In healthcare, in fact, we assist two different, and apparently irreconcilable phenomena (Kotler et al. 2010). On the one hand, there is a strong push to increase costs, both because

G. Turchetti (✉)
Istituto di Management, Scuola Superiore Sant'Anna, Pisa, Italy
e-mail: g.turchetti@sssup.it

E. Geisler
Stuart School of Business, Illinois Institute of Technology, Chicago, IL, USA
e-mail: geisler@stuart.iit.edu

L. Cinquini, A. Di Minin, R. Varaldo (eds.), *New Business Models and Value Creation: A Service Science Perspective.* Sxi 8, DOI 10.1007/978-88-470-2838-8_9, © Springer-Verlag Italia 2013

there is a growing demand for health services, due to the ageing of the population, the progress of medicine and a new concept of health – much more comprehensive in terms of attributes than it used to be – and there are persistent organisational and behavioural inefficiencies that cause overuse of emergency centres and hospitalisation. On the other hand, all governments, having serious concerns about the financial sustainability of the healthcare system, are introducing cost-containment measures and are working on how to reform the whole system (Turchetti et al. 2009, 2010, Geisler and Turchetti, 2011).

Even though it is difficult to find a correlation between the quality of the service and the cost of providing it,[1] in order to solve this difficult trade-off between higher demand for services and budget constraints, some authors have suggested increasing use of home care, promoting increased adoption of technologies (AMA 2001; Davies et al. 2009). Home care has long been recognised as a less expensive mode of provision of care (Hayes 2008), also because it reduces hospitalisation and the use of emergency centres. Some estimates suggest that the savings are in the range of 50–70% of in-patient costs for certain diseases such as chronic diabetes, cardio-pulmonary disease and arthritis (Kvedar et al. 2006; Geisler and Wickramasinghe 2009; Turchetti and Geisler 2010), at the same time guaranteeing high levels of quality of the services provided (continuous monitoring, appropriate treatments, increase of compliance, help to patient's family).

Home healthcare services are an example of services based on platforms,[2] as they rely on clinical and administrative technologies, generally known as "telemedicine". Telemedicine is an area of e-health defined by WHO as follows: "the delivery of healthcare services, where distance is a critical factor, by all healthcare professionals using information and communications technologies for the exchange of valid information for diagnosis, treatment and prevention of diseases and injuries, research and evaluation, and for the continuing education of healthcare providers, all in the interests of advancing the health of individuals and their communities".

In spite of their evident advantages, home healthcare and telemedicine are not very diffused yet. Why? This chapter addresses this issue by presenting and discussing an actual case of the use of wireless technologies in remoter home care for chronic patients.

Barriers to the implementation are listed and these are proposed as explanations of the very slow pace of the implementation process. In order to accelerate the pace of diffusion of home healthcare service, in this chapter we look at what Spoher and Kwan presented in the theoretic introduction to *service science* (see Chap. 1), and a service-dominant logic in the marketing of this technology-based service is therefore proposed.

---

[1] See Chap. 8.
[2] See Chap. 7.

9.2
Design and methodology of the study

The method used in this paper is an actual case of an experiment of using wireless technology in remote home care for chronic patients. We describe in this study the process by which wireless home care was designed to be introduced into the homes of a segment of the population of a large urban environment in the USA (Geisler and Wickramasinghe 2009). The study entails the participation of a large metropolitan trauma hospital. The population of patients was selected from the underserved and underinsured or uninsured patients with at least one chronic disease.

The problem facing the designers of the experiment was as follows. Underinsured or uninsured patients with chronic diseases such as diabetes, asthma, and pulmonary and cardiac illnesses tend to use the hospital's emergency department and trauma centre as their primary care facility. Since most of these patients lack health insurance and many are also undocumented immigrants, they prefer the emergency services of the hospital due to the fact that by law they are provided emergency care without any proof of insurance or citizenship. This behaviour of a relatively large segment of the population puts a financial burden on the hospital and severe constraints upon its availability of medical and administrative resources. The federal and state Medicaid (the programme that insures patients unable to pay or who lack insurance) program covers only a portion of, rather than the entire, bill for such healthcare services.

An analysis by the researchers and the hospital has concluded that the use of home care for this underserved population suffering from chronic diseases, particularly diabetes, will help to alleviate the financial burden imposed on the emergency resources of the hospital. A programme was established, together with a foreign company, to create a network of wireless users in the homes of these patients, linked to a central point of service in the hospital. Patients suffering from diabetes could be trained to use a specially designed cellular phone to dial into the hospital their daily reading of levels of glucose. A trained provider at the hospital would then determine whether any care was necessary and would advise the patient, by telephone, whether the patient should come to the hospital or help would be provided in the patient's residence. This process would also allow caregivers to dispense with the traditional practice of having patients keep a journal/diary in which they would annotate their daily readings of levels of glucose. Daily entries into the cellular phone would provide the hospital with immediate electronic data on the patient's condition as well as weekly and monthly trends. Many patients would traditionally come to the hospital, particularly in a state of urgency, and would forget their journal or forget to enter readings of certain days when they felt better. The resulting electronic database would also allow caregivers to have a sample of patients that could be used for research and statistical analyses.

The experimental study was put into operation in early 2008 and in the first stage the hospital and the private company agreed on the hardware and the software to be used in the study. A sample of 12 patients was selected and the next stage was the marketing of the procedure to the patients. Detailed face-to-face interviews were conducted with patients, their families and hospital professionals (both from the medical and administrative areas). Similar experiments were conducted in other urban settings in the USA and Canada, where the samples included wealthier groups of patients with diabetes. This experiment was more successful. In both cases the patients expressed a high level of satisfaction (Wickramasinghe and Goldberg 2004).

In the case described here, marketing the home care services via the use of wireless technology became an impediment to the implementation of the procedure. There were several categories of barriers that impinged upon the successful marketing of the wireless technology to the sample of patients and their relatives.

9.2.1
Barriers to marketing

Four distinct categories of barriers to marketing have been identified. They contained several factors acting as impediments to the successful marketing to patients.

1) **Technology.** The key barriers in this category were the set-up and training of the patients and the hospital staff, and the connectivity issues. Although many patients and their relatives were accustomed to the use of cellular phones, they were reluctant to learn the procedure needed to send and receive data from the modified phones. Issues of connectivity also arose, particularly within the hospital, in which there was a need to establish an independent centre for data collection from the home phones but also to link this centre to other parts of the hospital (Turchetti and Geisler 2010).

2) **Human Behaviour.** Perhaps the strongest set of barriers, this category includes such impediments as: fear of innovation, distrust and data privacy. The underinsured and unsophisticated patients exhibited an innate fear of innovation and new technologies. They also feared that the cellular phones were another form of government intrusion into their lives and another means for the government to keep constant control over their movement. Although caregivers carefully explained the specific uses of the phones for the sole collection of clinical data, the fear of allowing such technology into their homes was paramount on the minds of the patients and their families. They also feared the possible misuse of the medical data they provided for purposes other than medical care. The patients and their families explained that with a written diary they carry to the emergency department of the hospital, they had control over who received and read and used their medical information. With the electronic provision of medical information, the patients and their families lost control over where the information was going and who would use it, and for what purpose. Patients with diabetes who had employment feared that their employers might be provided with this information and decide to terminate their employment (Berry 2006).

3) **Organisation.** Caregivers experienced issues of resistance to change and the need for special training and special skills to handle the new electronic database of patients with diabetes. Since most of the caregivers at the hospital were already overburdened with short-term problems and scare human resources, the

special training and added responsibilities were met with scepticism and lack of enthusiasm (Bevan and Robinson 2005).

4) **Economic.** The economic barriers were primarily in terms of the cost of the procedure and the inability of the experimenters to clearly show cost-savings or the benefits to patients that would compensate for the added costs of establishing the process, training the staff, selling to the patients and maintaining the process.

9.2.2
Marketing to patients and providers of healthcare services

The marketing effort to patients and providers was designed to overcome the barriers listed above. The experimenters used three tactical avenues. The first was to fully explain the procedure and to minimise the onus of the set-up and the training. The second was to fully explain the potential benefits to the patients, their families and the hospital. The third was to fully explain the destiny of the information, the secrecy that would be maintained and the absolute guarantee that the information would never be shared with any government agency or office, and that any information used for research will be in the aggregate only and would not have any of the individual names (Earp and Payton 2006).

The marketing effort directed at the patients seemed to be only partially success-ful. Few patients and their families accepted the explanations of the nature of the process and the issues of data privacy. Even when caregivers joined the marketing effort (to increase the level of trust of the patients), the vast majority of patients resisted the idea that they should introduce the new wireless technology into their homes and into their lives. Some also expressed the strong fear that if they would send their information electronically, the hospital would then totally dispense with its services and would not allow them to use the services of the emergency depart-ment of the hospital. They would not accept the premise that the cellular phones were not a replacement for actual care at the hospital but simply a more efficient substitute for the written journal they traditionally used to record their glucose readings.

Although the logic of the home care service was clear and dominant, it was an unacceptable alternative for the patients (Vargo and Lusch 2008b). It was hardly enough for the experimenters and the hospital staff to enumerate the advantages of the new technology and its benefits – patients and their families would not accept the service. Fear, ignorance and cultural factors – all quite understandable – have acted in concert to impede the marketing of these wireless technologies and the service they were meant to provide. The marketing effort described here had a duration of about six months, with a cost of about 1800 human hours. The experiment is still in progress. The hospital and the experimenters are still hopeful they can overcome the resistance of the patients and their families.

The continuation of the experiment will certainly require additional resources and perhaps a different approach to the marketing of the services to the patients (Vargo and Lusch 2008a). The enumeration of the attributes and benefits of the ser-vice seems insufficient to convince the patients and their families (the customers of the service) that the service and the technology are a worthwhile investment and that their involvement would be beneficial to them and to the management of their illness.

9.3
Findings

The findings from this study can be clustered into two major categories. The first is the conclusion that technological superiority alone is not enough to persuade customers to buy into a product or in this case a service. The second category is the conclusion that *service-dominant logic* can be a positive force in persuading customers to implement technology if it addresses the issues and barriers exhibited by these customers.

9.3.1
"If you build it, they will not necessarily come"

The notion that if a technology is adequate or superior and if its attributes and effectiveness can be demonstrated, then customers will come/buy it, is not necessarily true. In this study we found that the barriers to implementation of the technology – hence to the acceptance by the customers of the service – are not technological but are anchored in the behavioural issues of the patients/customers. Even a superior technology, already proven in other instances, cannot be marketed to certain segments of a target population unless the marketing effort addresses the concerns, barriers, fears and uncertainties of that population. The logic of the service and its appropriateness and potential benefits are unacceptable to certain customers. In healthcare delivery it is the patients who are viewed as customers of a service which, by definition, is designed to help them and to provide them with relief and medical care. Even in this case, where care is the service, there are non-technological barriers strong enough to prevent the service from being marketed (Turchetti and Geisler 2010; Kotler et al. 2010).

Technology alone is not sufficient to market a service. As shown here, even in the case of healthcare delivery, the service itself – however effective and attractive – is not enough to successfully market it to the target population. There is a need for an "extra dose of logic", tailored to their needs, which will provide the target customers with a description of the service aimed at the challenges of the barriers that guide their logic and their purchasing behaviour.

9.3.2
Service-dominant logic

Concentrating on the logic of providing home care services is the preferred way in which such services can be successfully marketed to a target population of patients. Although the initial effort at marketing such technologies has been unsuccessful, the conclusion from this study is that there is a need to restructure the marketing effort and to tailor the logic of the service to the challenges of barriers to implementation and to marketing (Vargo and Lusch 2008b). More specifically, the conclusion is that there is a need to design the marketing effort around a carefully structured *service-dominant logic* which addresses each and every barrier identified in this study. It is not enough to provide a general logic of potential benefits and the attractive attributes of the technology and the procedure. The marketing campaign must painstakingly list each and every barrier and address them with careful consideration. The marketing campaign must also enhance the role that facilitating factors have in convincing patients to accept the technology-based services, but these factors are less powerful in their ability to persuade customers or to allay their fears and uncertainties.

There are at least two ways or marketing tactics to accomplish this purpose. The first is to clearly list the barrier and to declare that the service will or will not cause this to happen. For example, if the barrier is the fear of the misuse of the information provided by the patient, the marketing effort declares that the information will be strictly used for clinical purposes only. The second way is to provide the patient with examples or cases of similar applications to similar categories of patients. In the example above it is useful to provide examples of the uses of other medical data received from underserved patients and the appropriate ethical use of that data in those cases.

In order to make the marketing effort effective, both technology producers and hospitals/healthcare services providers have to change their approach, adopting a *service-dominant logic*. From the one side, the technology producers spoke only to the healthcare services providers, avoiding contact with the patients and their families, and from the other side, the healthcare services providers – even though they are, as the patients are, the customers of the technology – played the role of the sellers towards the patient in order to reach their objectives (reduce costs, hospitalisation and use of emergency centres). In other words, in the presented case study, we observed two "negative" approaches: (1) the producer did not adopt a *service-dominant logic* approach; (2) one of the customers/users, the hospital, behaved as a producer, not adopting a *service-dominant logic* approach itself, thus failing in cocreating value both with the producer and the other customers (see Fig. 9.1).

The conclusion from this study is that although the initial marketing effort was unsuccessful, it succeeded in identifying the barriers that impeded the acceptance of the technology and the service by the target population of patients (Geisler and Wickramasinghe 2009). This accomplishment was sufficient for the experimenters and the hospital to allow for the restructuring of the marketing programme, and in this sense it was a success.

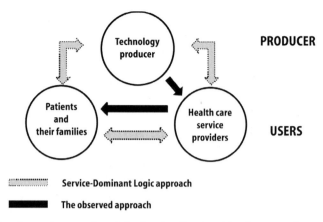

Service-Dominant Logic approach

The observed approach

Fig. 9.1 The observed approach and the Service-Dominant Logic approach

9.4
Managerial implications

This study has several practical implications for service industries. This chapter dealt with the implementation of home care and telemedicine. Marketing of remote care for the home has proven to be a difficult problem, requiring a modified marketing scheme that addresses the barriers to implementation. These are barriers also common to traditional implementation and adoption of technology, particularly information and telecommunication technologies.

Four categories of practical implications for the service industries, transportation, finance, hospitality, and healthcare, are identified. Although each industry sector has some specific needs and requires a somewhat differentiated approach to marketing of technology-based services, the implications listed in this chapter apply across all of these industrial sectors. Once an innovation path of the business model has been defined,[3] the following strategic issues have to be taken into consideration as they are particularly relevant for the marketing of a service.

9.4.1
"One size doesn't fit all"

The notion that the marketing of technology is a uniform activity across service sectors is wrong. The practical implication from the case study reported here is the need for a tailored marketing approach for each sector. In healthcare there is a need to identify not only the attributes and capabilities of the technology being marketed, but also the special needs and factors inherent in the potential customers/patients.

---

[3] See Chap. 3.

Marketing of technology for the healthcare sector requires tailoring the marketing approach to specific barriers and addressing each one of these barriers.

In the financial sector, for example, the practical implication is embedded in the need to market information and telecommunication technology in a manner that addresses the constraints and the specific barriers inherent in the industry and in the customers. Too often vendors of information and telecommunication technologies employ a standardised marketing plan for all service industries. "What works for transportation will also work for banks and for hospitals." The initial implementation of information technologies in the banking industry in the USA failed mainly because vendors looked for a unified model of an all-encompassing network of banks. They underestimated the complexity of the banking system and the unique attributes of individual banks, their specific needs and the unique barriers they had.

9.4.2
"Technology is not enough"

Information and telecommunication technologies, however exciting and effective, do not sell themselves. The practical implication for other service sectors is that technology alone is not enough to successfully market. In the transportation sector, for example, the existence of effective information technology for the sale of airline tickets was not enough for all airlines to implement the technology. With the evolution of the Internet, the technology, first developed by American Airlines as a workable ticketing system, acquired a successful path of adoption, mainly because the Internet resolved many of the problems and addressed many of the barriers to implementation. Unlike traditional products, such as appliances or automobiles, the implementation and adoption of technology for the service sector requires the additional employment of service, rather than *technology-dominant logic*.

9.4.3
"If at first you don't succeed, try, try again"

Another practical implication is the notion gained from this case study that an initial attempt to market the technology to a service industry may fail, but the failure is a blessing in disguise. The failure may help to identify the barriers inherent in the service industry and its customers. This apparent failure allows the marketers of technology to restructure their marketing approach and tactics to address these barriers. The initial failure of the implementation of information and telecommunication technologies in the financial sectors has led to the revision of the marketing strategy and tactics of key vendors. These companies reassessed their approach and restructured their marketing effort so as to address the barriers that they had uncovered during the failed effort.

9.4.4
"From propaganda to conversation"

In the presented case study, the service has been marketed as a good, showing and communicating the advantages for the patient and the hospital. Neither the patient and his/her family nor the hospital professionals (administrative and medical) were adequately listened to and involved in the process of creation of value. For this reason they continued to see the service from the outside, maintaining their doubts and aversion to the proposed opportunity. The benefits and the creation of value related to the service were not clear to the users. They were told what they had to do (provide information and data) and as a result they did not perceive the benefits, the value for them. On the contrary, they believed that the home care service would only benefit the hospital (which would mean a decrease in hospitalisation and use of emergency centres) and reduce their opportunities and level of service (they felt that adopting the home care service would limit their chance of receiving face-to-face services and accessibility to healthcare structures).

Therefore, as happens for other service industries such as financial services and the professions (lawyers, engineers, architects, etc.), it is necessary to bring the customer (in our case study, the patient, his/her family and the healthcare services providers) and all the other relevant stakeholders into the process of creation of value, talking with them – not simply informing them – from the very beginning of the process. It must become a conversation and a cocreation of value (Lusch and Vargo 2008; Payne et al. 2008).

9.5
Originality and value of the study

This case study provides some evidence of the proposition that the main barriers to the marketing and adoption of telemedicine in home care situations are organisational, behavioural and economic, not technological. This evidence is in line with the traditional findings in the marketing of technology in other areas, such as industrial research and development, and the marketing of new *technology-based* products by industrial companies.

The findings from the case study described in this paper offer an explanation of the initial failure of the marketing effort aimed at the implementation of remote care wireless technology for certain segments of patients with chronic diseases. The provision of home care is an example of a service *mini-world*, encapsulated within a critical service industry of the delivery of healthcare. The value of the case study described in this paper is primarily in findings that explain a paradoxical phenomenon of the rejection of improved healthcare delivery due to the issues associated with the adoption of new technologies. Underserved patients with debilitating chronic diseases resisted the implementation of improved services and failed to accept the logic of the potential benefits from these improved care services. Instead of embracing such improvements, which would allow them to stay in their home and be monitored from the hospital via the use of cellular phones, the patients and their families resisted the change and rejected the new technologies.

The implication for other industries in the service sector is that if customers of the healthcare industry reject services due to their resistance to technological change, it stands to reason that in other service industries such as transportation, hospitality, financial services, education and the professions (lawyers, engineers, architects, etc.), in which *technological-based* services are marketed that are less critical than healthcare, resistance to technological change may have an even higher impact on the nature and the success of marketing of these services. The value of the findings from the case study reported in this paper is therefore in the lessons learned from the case study in healthcare services and the extension to other industries in the service economy. The provision of services alone (even at no cost to the customers, as is the case with the underserved population of patients described in this case) and the existence of a sound logic to the marketing of such services are not sufficient to successfully market the services to prospective customers.

References

American Medical Association (2001) American Medical Association guide to home care. Wiley, New York

Berry L (2006) Creating new markets through service innovation. MIT Sloan Manag Rev 47:56–62

Bevan G, Robinson R (2005) The interplay between economic and political logics: path dependency in health care in England. J Health Pol Policy Law 30:53–78

Davies S, Froggatt K, Meyer J (eds) (2009) Understanding home care: a research and development perspective. Jessica Kingsley Publishers, London, UK

Earp J, Payton F (2006) Information privacy in the service sector: an exploratory study of health care and banking professionals. J Organ Comput Electron Commerce 16:105–122

Geisler E, Turchetti G (eds) (2011) Management of technology in healthcare organizations. Int J Healthc Technol Manag 12(3/4):195–349

Geisler E, Wickramasinghe N (2009) The role and use of wireless technology in the management and monitoring of chronic diseases. IBM Center for the Business of Government, Washington, D.C.

Hayes H (2008) Home health monitoring saves the government big bucks. Available at: http://www.govhealthit.com

Kotler P, Shalowitz J, Stevens RJ, Turchetti G (2010) Marketing per la Sanità. Logiche e Strumenti, McGraw-Hill, Milan

Kvedar J, Wootton R, Dimnick S (eds) (2006) Home telehealth: connecting care within the community. Royal Society of Medicine Press, London, UK

Lusch RF, Vargo SL (2008) The service-dominant mindset. In: Hefley B, Murphy W (eds) Service science, management and engineering. Springer Science+Business Media, LLC

Payne AF, Storbacka K, Frow P (2008) Managing the co-creation of value. J Acad Mark Sci 36:83–96

Turchetti G, Geisler E (2010) Economic and organizational factors in the future of tele-
    medicine and home care. In: Coronato A, DiPietro G (eds) Pervasive and smart tech-
    nologies for healthcare. IGI Global, Hershey, PA

Turchetti G, Krabbendam K, Geisler E, Wickramasinghe N (eds) (2009) Key considera-
    tions in responsible healthcare technology management. Int J Healthc Technol Manag
    10(4/5):223–359

Turchetti G, Spadoni E, Geisler E (2010) Health technology assessment. Evaluation of
    biomedical innovative technologies. IEEE Eng Med Biol Mag 29(3):70–76

Vargo S, Lusch R (2008a) Why Service? J Acad Mark Sci 36:25–38

Vargo S, Lusch R (2008b) Service-dominant logic: continuing the evolution. J Acad Mark
    Sci 36:1–10

Wickramasinghe N, Goldberg S (2004) How M=EC2 in health care. Int J Mobile Commun
    2:140–156

INNOVATION LAB
From contactless to multiservice cards

MAINS Master, academic year 2009/2010
People and companies involved in the InnoLab:
Students: Massimo Di Stefano, Alessia Innocenti and Marco Rosabella
Companies: Poste Italiane, Telecom Italia and SIA-SSB
Professors: Roberto Barontini and Giuseppe Turchetti

1. The problem
In a scenario that looks at today's consumers as users of a variety of services, the work aimed to develop an open multi-service platform that embraces different services of wide interest.

The world of services in fact provides a rather fragmented offer, which translates into an ample series of support activities and infrastructures: the plurality of interfaces renders desired services that are not very usable and are not always in step with technological evolution. This work, aiming to exploit the ever-growing contactless technology, is positioned within the service digitisation trend.

The first problem addressed was the identification of levers to create the need in the consumer to adhere to the proposed multi-service platform, which is closely linked to the choice of services to be included and the technology identified for its fruition.

2. Work methodology
The project was developed in several phases with different purposes.

The study began with scouting of existing standards and the use of contactless technology in various fields, in turn identifying critical factors and those that potentially delay its diffusion.

Thus, the team selected a range of services with the ultimate goal of identifying the best way to implement this service and make it easily accessible.

The first phase of the work foresaw the study and research of innovative solutions already on the market, both in technological terms and in terms of implemented services/multi-services to be applied to the pilot project.

From the services perspective, those currently most widespread using contactless technology were analysed, in particular in the transport, mass catering and payment sectors in different countries, identifying critical issues that bring about delayed diffusion.

This screening allowed the choice of only those services that are technically and economically feasible. Three critical services were identified: meal vouchers, museum tickets and loyalty cards for large retailer chains.

3. Proposed solution

The analysis performed led to the "meal voucher" service, used in Italian companies, as a killer application of the entire project. This service was chosen after careful market analysis showed that this service is established and growing fast, with sales increasing by 7.6% over the past 5 years and concentrated on a few actors (the top three cover around 60% of the market).

Contactless technology was chosen for the number of advantages offered including payment convenience and speed. More specifically, the choice of this technology was applied to the world of cellular telephony, namely near field communication (NFC).

NFC is a short-range two-way communication technology that allows two devices in close contact to exchange data within a radius of 2–5 cm. Although mobile phones with integrated NFC exist, their diffusion is still slow. However, it is possible to add the contactless functionality to mobile phones currently in use through plug-ins such as "NFC stickers".

This is a simple sticker to be applied to mobile phones enabling two devices to communicate using Bluetooth technology. An application relating to the desired service is installed in the device's flash memory while the built-in antenna allows communication with NFC readers during each transaction.

The strategy envisaged addressing a market segment initially consisting of the employees of companies that join the "meal voucher" supply circuit. Currently, the meal voucher issuer stipulates a supply contract with a company providing customisation and delivery of paper meal vouchers. The company then deals with the distribution of monthly meal vouchers to employees who use them at participating outlets. In turn, the outlets collect the meal vouchers and, at the end of each month, send them to the issuer for reimbursement as and when foreseen.

With the introduction of the sticker as a means to access the service, the entire system changes as follows: A supply contract between the company and meal voucher issuer remains, but the vouchers are directly loaded onto the stickers of company employees, thus eliminating all the paperwork and logistics costs. The validation of meal vouchers at participating outlets comes about electronically, as does sending the vouchers collected by participating outlets and their subsequent reimbursement.

Within this circuit, the advantages over the existing supply system are numerous and concern all players involved. The service provider attends to the purchase of stickers and customisation, ensuring the functioning of the new infrastructure in the face of a hypothesised margin of 2% on profits. The meal voucher issuer, who already has the circuit of participating outlets in place, will achieve lower operating costs in the face of a profit reduction of 2%. The company will achieve lower management costs and pays only for meal vouchers actually used. The outlet will see a reduction in the reimbursement delay as a result of full automation in addition to a reduction of queues due to the speed of transactions. Finally, employees can more easily take advantage of meal vouchers via their mobile phones, a devise that is used daily, but will moreover own a medium that will allow them to join a multi-service platform, which is the true added value of the entire project.

In this perspective, the work led to the identification of two other services to be added to "meal vouchers" in the multi-service platform pilot project: a "loyalty" service, widely used in various areas including large-scale retail chains, and a "ticketing" service for museums.

These two services were chosen since the former is the most widely used loyalty retention tool and is also a way to gather information on consumer purchasing behaviour. Moreover, its use is rapidly growing and is widely popular among consumers, rendering the platform attractive to large-scale retail chains. The choice of museums should be seen within a wider web ticketing service, which is not widely used but if incentivised could benefit both consumers and entities operating in the culture and entertainment sector.

Finally, the team evaluated strategies to increase the reliability offered by the platform in terms of security. The best solution identified provides a centralised database on which to back up data, enabling the multi-service of each sticker so that this becomes a simple access point to the database.

A feasibility study was carried out in economic terms in support of the decisions. The evaluations confirmed the benefits and the sustainability of the project. In particular, assuming a constant growth in sticker sales up to 1 million in the fifth year, profits would already be achieved from the third year onwards. In addition, total sticker sales of just under 400,000 units would also be sufficient to recover the investment.

An essential impetus for the success of this innovative tool to use more services will come from the spread of mobile phones with NFC technology, which would replace the current sticker interface. Finally, the achievement of critical mass will ensure that this platform establishes itself as the reference standard among consumers and expands among customers as the number of services included increases.

The management and governance of new service models in the environmental and energy sectors

10

Marco Frey and Francesco Rizzi

Energy and environmental services are affected by a particularly significant dynamism correlated to green-oriented innovations. Although the so-called "green economy" cannot yet be considered as a new techno-economic paradigm, it constitutes a complement that can enrich and stimulate the innovative trajectories in green services beyond the contribution of the evolution of ICTs. The green economy does not entail only contributions from renewable energy as alternatives to fossil fuels but also a thrust towards sustainability and efficiency that is not limited to industrial sectors (e.g., consider the significant advances in energy–environmental innovations in relation to the construction industry). Greening the economy is possible through an approach oriented to the integration of cycles of those natural resources whose value increasingly affects overall systemic performances. This integration is fundamental both at the policy level (i.e., regulation and control at national but also regional level), at the service provision level (e.g., integrated waste management, water management, etc.) and at the corporate level (e.g., resource productivity). In this context, business models based on process and service innovation are strengthening from the perspective of interrelated production chains where the role played by users is increasingly central (in saving energy or water, in waste collection, etc.). Accordingly, enterprises are consolidating approaches based on Life Cycle Assessments, on integrated management systems (where specific energy management standards are designed to complement environmental management standards) and on eco- and energy efficiency.

M. Frey (✉)
Istituto di Management, Scuola Superiore Sant'Anna, Pisa, Italy
e-mail: m.frey@sssup.it

F. Rizzi
Istituto di Management, Scuola Superiore Sant'Anna, Pisa, Italy
e-mail: f.rizzi@sssup.it

L. Cinquini, A. Di Minin, R. Varaldo (eds.), *New Business Models and Value Creation: A Service Science Perspective.* Sxi 8, DOI 10.1007/978-88-470-2838-8_10, © Springer-Verlag Italia 2013

10.1
Environmental and energy services, green economy and system innovation

In 1994, the International Council for Local Environmental Initiatives (ICLEI) provided one of the first definitions of sustainable development, after the classic Brundtland definition (WCED 1987), portraying it as "development that delivers basic environmental, social and economic services to all residences of a community without threatening the viability of natural, built and social systems upon which the delivery of those systems depends" (ICLEI 1994). Although as time has passed this has been surpassed by more extensive definitions,[1] this formulation has the merit of recognising to environmental services a function of fundamental importance in maintaining the environmental, economic and social equilibrium of a community. Since these three dimensions of sustainable development are closely interrelated, the debate concerning this equilibrium has helped spread awareness of how all interventions on service programming must take into account mutual interrelations, thus implicitly exceeding the technocratic vision that circumscribed the scope of their evaluation on only the punctual parameters of industrial performance.

In light of this and subsequent efforts to delineate the correct boundaries of the regulation of environmental services (Brusco et al. 1995), a perimeter based on the concept of service life cycle was consolidated, according to which the following are attributable to "primary collective services" of an environmental nature:

1) management of the integrated water cycle in its collection, treatment, distribution and purification stages;
2) management of the integrated physical resources/waste cycle intended from its conferment, collection, transportation, treatment disposal and reuse stages;
3) management of the energy chain, in turn divided into the stages of production (or generation), transmission, distribution and processing.

These areas, although divided by the particularities of their technological and managerial challenges, are now united by the same needs to reorganise and adapt to changes in society.

Among the interpretations of this trend, new consumption and access to service models – collocated among the main impetuses capable of generating growing demand for technical and management innovation – increasingly lead to indicating the so-called green economy as a unifying opportunity to redefine business models and the integration of services across and between sectors.

Although the concept of the green economy has to date tended to primarily emphasise the reorientation of the energy sector with respect to the global warming challenge (low-carbon economy), this perspective is pervasively opening up towards the transition of the economy towards sustainable development. In fact, in this arena, equal importance with respect to the energy cycle is assumed by other cycles, such as water, waste and the agro-industrial chain, designed to provide other important

---

[1] For example, UNESCO (2001) broadened the concept of sustainable development, indicating that "cultural diversity is as necessary for humankind as biodiversity is for nature".

Fig. 10.1   The systemic nature of the Green Economy (Symbola 2010)

investment and innovation contexts. From this perspective, those products and services that ensure low environmental impact throughout all stages of their life become competitive.

Further integration into supply chains has come about whereby the two-way relationship between manufacturer and customer is replaced by an open relationship involving the various actors of the entire system. This evolution leads to the concept of shared responsibility where designers, manufacturers, distributors, end users, but also institutions and citizens are not separate parts of a linear path, but interdependent actors in a dynamic and complex system of relationships.

The definition of green economy that we intend to adopt is thus not only focused on the business opportunity offered by new technology and techniques in response to the emerging scarcity of resources (of energy, water, food, resilience to greenhouse gas emissions, etc.), in a perspective that we may briefly define as green business, but also on the possibilities linked to an evolved economic system where the business offer is accompanied by a consumer demand aware of sustainability issues, by the responsible behaviour of citizens and by the policies pursued by institutions capable of looking at the long term.

This holistic and systemic approach (Fig. 10.1) is exploited on both the global and local levels, where the quality of services is better able to influence performance in terms of quality of life, as well as the attractiveness of economic activities and human capital.

The complexity of the nature of the public–private service, which is an indispensable determinant of its organisational structure, rises from its diffusion in the territory.

Traditionally, in the production sector, the service concept is associated with activities aimed at creating complementary value and, even more, the retention of specific customer segments. Therefore, every technological and management innovation activity here has the primary objective of diversification with respect to the competition and the interception of new market segments.

In the environmental and energy services industry, the relationship with the customer, necessarily passing through an important process of creating the physical infrastructure of collective interest, instead becomes not merely a structurally durable element but also characteristic of a territory and, therefore, the terrain on which to assess its governance. This is not only a perfect reversal of the perspective of interest from the interception of demand to guarantee the offer, but universal access and sustainability of the service are introduced alongside the fundamental principles of efficiency, effectiveness and economic viability.

The requirements for strict standardisation in the design, implementation and management stages that can be found here thus need to be counterbalanced by the importance of the socio-economic, environmental and technological contextualisation. Hence, for example, municipal waste collection methods should take into account the socio-cultural and economic fabric that they serve, drinking water distribution must consider the quantitative–qualitative state of tapped sources, and energy supply the sources and sinks in place in the network.

The sustainability of the management of local resources plays a role, on one hand, in the capacity to control the evolution of such relationships and, on the other, in the elimination of so-called lock-in[2] in future innovations.

Among innovation processes, furthermore, those relating to the development of professionals should not be underestimated. In fact, increased training needs can be identified in both the structures responsible for regulation as well as in the people called on to work in different parts of the production chain. New professional skills of an interdisciplinary nature must be created, for example, to operate effective communication in the early stages of plant development, to coordinate complex financial management in the event of recourse to project financing, to move on voluntary emissions exchange markets, to document the sustainability of own offers, etc.

We can glimpse in the progressive projection of customer satisfaction of the management of formal service aspects a surpassing of the "commodisation"[3] processes of water, energy and waste management.[4] If, in fact, formerly, the physical dimension of the product was perceived as less valuable with respect to the possibility of being or not being regularly available, then today new visibility of the objective of the service is given by the recognised impact that such services have on the social system (given the multiple uses, as is the case with water, or – more generally – the correlation with the quality of life), the economic system (given the increasing cost of services that were once almost free) and the environmental system (with consequent implications in terms of social acceptability). It is here, for

---

[2] See Liebowitz and Margolis (1994).

[3] A concept derived from the term "commodity" that indicates a good that has demand but is fungible and offered on the market with no qualitative differences, or rather, the product is the same regardless of who produces it.

[4] Picazo-Tadeo et al. (2008), analysing Spanish water utilities, observe that the measurement of water quality influences the performance of the different parts of the water service supply chain.

example, that the evolution of the concept of waste management finds justification, namely from "refuse collection", a maintenance service of the public good, to "local public service", which is able to impact on local competitiveness. In other words, a service that has the objective of "the production of goods and activities designed to achieve social aims and to promote the economic and social development of local communities".[5]

The fact that the operator's target is positioned along the territory–consumer axis also creates interesting potential diversification scenarios and, therefore, new possibilities to compete on different alternatives of the offer.

The relevance of this issue is demonstrated by the fact that in the recent past, liberalisation policies of public utilities have focused precisely on the differentiation of potential service providers. In many cases, these policies have led to a common process of privatisation, but with different results in terms of interpretation and penetration of the competition. These are in fact processes that involve huge amounts of capital and networks of competencies, capable of generating bottlenecks on both the supply side (e.g., the willingness of new entrants to take over previous management) and on the demand side (e.g., the willingness of consumers to bear the costs of the improvement of guaranteed services).[6]

To understand how these aspects are able to affect the trajectories of system innovation requires recollecting the fundamental link between the reform principles and the economic and financial management configuration to be activated. In fact, the principle of Full Cost Recovery inspires the privatisation process or, rather, economic and financial management autonomy and separation from those interventions in the general fiscal system frequently required to reconcile industrial costs with the level of tariff promised to citizens by local administrations. This is how "controlled" liberalisation was promoted according to European dictates on services of general interest: namely, a gradual opening of the market, accompanied by measures to ensure protection of general interest in order to guarantee, in line with the concept of universal service, access for all to a service of specified quality and at an affordable price, regardless of the economic, social or geographical situation.

In this context, wide discussion has ensued on what the possible variations of the operational concept of service of general interest are, as well as public service obligations arising therefrom. This is how an increasing number of interpretations have come about of the roles that public and private actors should have, taking into account the specific circumstances of each sector, in securing universality and equality of access from a technical and managerial profile as well as continuity, security, adaptability, quality, efficiency, affordability, transparency, protection of disadvantaged social groups, the protection of users, consumers and the environment, and the participation of citizens.

---

[5] Definition of local public service introduced by Decree 267/2000 "Uniform laws on the organization of local authorities".

[6] Warner and Bel (2008), comparing the management of some public services in the USA and Spain, conclude that the free market cannot be assumed *a priori* as a solution for the most efficient provision of local services, revealing that at times the public role is a useful factor in regularising the offer.

All this has helped, according to a popular conservative approach, defend the positions that can immediately guarantee achieving minimum targets in terms of service provision.

Theory underlines that innovation dynamics are usually emphasised in systems characterised by a "related variety" of competencies and actors involved in finding the best technical management compromise (Bishop 2008). Unfortunately, the importance of investments in the above-mentioned sectors and the consequent critical mass required to ensure the service provision have imposed constraints and significant entry barriers for new players and for the penetration of innovative ideas. Additionally, the aggregation of different classes of demand in a single standardised volume, through the processes of homogenisation of their distinctive features, has in many cases prevailed and helped to marginalise innovation efforts in user niches, focusing resources on ensuring a minimum functional level of infrastructurisation of the territory.

On the territory-customer-service provider axis, not only the game of the above-mentioned general human, economic and technical resources is played, but also the coordination of these with the specific environmental resources available locally. Thus, the socio-cultural dimension often becomes, for example through the dynamic creation of citizen consent, a possible explanatory variable in the success or failure of technological innovation promoted by the operator in terms of actual environmental impact on the territory.

A typical example of this is the introduction of capillary systems to measure levels of service. Often judged as fundamental to better planning and achieving efficiency of the service provision, the adoption of these systems depends on the expected impact on consumer behaviour: for example, remote sensing tools that can bring benefits to balancing water and electricity distribution pipelines and reduce the operator's financial exposure,[7] are extremely damaging to the perception of risk of inducing irregular and improper conduct in the case of waste collection.[8]

The introduction of technical and managerial innovations in every part of the chain is configured as the result of complex choices, subject to subsequent reconsideration in light of the evolving nature that characterises the management of those systems that are, *de facto*, learning organisations (Miles 2006).

In light of the above, even so-called "green" investments, i.e., those that combine environmental and economic objectives through more efficient use of resources, more often find room through imitation mechanisms rather than competitive advancement. Given the nature of the problems and priorities to be addressed in this first system redesign phase (e.g., the size of the smallest management units and cap-

---

[7] For example, through more frequent and proper billing of volumes consumed.

[8] For example, a measurement of actual waste production of each single use is dreaded by those who, referring to experiences in the management of special waste, believe that this could be an incentive to resort to illegal forms of disposal. If therefore, on the one hand, the pay-as-you-throw based tariff would foster competition between disposal alternatives, stimulating the introduction of an alternative valorisation of waste materials at a lower cost to the producer, on the other, it would introduce the alternative of resorting to illegal disposal channels.

ital at stake), the introduction of incremental innovations thus largely prevails on radical innovations.

With particular reference to water supply and urban hygiene, the key challenges now concern access to existing technologies rather than their innovation. For example, in the water sector, evolution over time has led to the emergence of numerous natural monopolies on a local scale (consistent with sub-river basins). Here, services are typically focused only on the most critical cost items (basically the expansion of networks and their maintenance in order to contain losses), resulting in the dominance of low-tech interventions. Moreover, in these segments of the chain, the traditional presence of mature technologies has not imposed specific requirements in terms of technology transfer. Nevertheless, since investment in new technologies has not been included among the critical factors of the system's evolution, the judiciousness of optimally investing revenues derived from purification and distribution services throughout the chain has failed. It is no coincidence that, in order to reduce the most significant operating costs (including those related to energy linked to the pumping of water and the disposal of sewage sludge), the demand is multiplying for technologies that, already available for several years, may now be recognised as profitable (e.g., low-power sections for remote monitoring of the network).

Precisely in this sense, the focus on the green economy could provide a decisive impetus to a cultural revolution that, reassigning the correct importance to marginal and auxiliary processes and to the creation of knowledge, could lead to significant financial and environmental results through a punctual and systematic introduction of incremental innovations. Such a process, stimulated by a new generation of environmental policies, amplified by the increasing attention to cost optimisation imposed by the refinement of the privatisation processes, is the conceptual framework in which to seek the optimisation of resources in view of the entire life cycle of products/services.

10.2
The model of integrated cycles at macro, meso and micro levels

The characteristics of the environmental services described above enable identification of three levels of integration of their determinants:

- A first level, of a technical nature, concerns the provision of services from the perspective of an intra- and inter-production chain life cycle.
- A second level, of a management nature, is ascribable to the dimension that captures the highest possible corporate intra- and inter-synergies from an economic profile (see compensations in the cost chain) and programming efforts are expressed more efficiently.
- A third level, linked to policies, is determined by the need to coordinate regulation activities at central and local levels, thus ensuring the balanced development

of the chains, the subsidiarity between territories and the governance of cross-media effects on a local and global scale.

In analysing these levels of integration, it is thus crucial to grasp their complexity not only through different sectoral perspectives, but also through different observation scales:

- the macro scale, describing how the above-described determinants do or do not find the right equilibrium in the system (e.g., how investments in material recovery are coherent with energy and product policies);
- the meso scale, where the relations between actors called on to implement the system, and the relationship between geographical areas and local policy makers become important (including, for example, the resolution of the NIMBY[9] syndrome of public infrastructures);
- the micro scale, which includes the executional capacity of individual actors (firms, government authorities, research centres, social communities, etc.) called on to manage local sustainable development.

In this threefold perspective, the different layers of policies of the EU design can be recognised in:

- the national layer, focused on trade and material balances and on regulation;
- the regional layer, based on planning and infrastructural autonomy;
- the local layer, centred on the compatibility of services with the delicate equilibrium between local expectations and needs, including resources and constraints.

The framework is developed along these levels to correctly interpret the transition from the "service obligation logic", as was the case with public administrations in the past, to the "market logic" of the operating firms.

The chain of coordinated objectives that from the macro level move towards decisions on the micro scale, typically opens up to the legitimacy of two possible schemas that can be activated at the local level: a dual system where the leading role is attributed to the party assigned with overseeing the operational side, and an integrated model, with the shared responsibility of different parties in the design and management of the entire supply chain.

The prevalence of one model over the other is determined by the conditions imposed locally and, more importantly, the possibility of ascribing these to common management within homogeneous territories. The concept of integration in its various forms, given the continued supremacy of economies of scale over those of scope, is emerging as a key strategy to achieve operational efficiency.

In the waste collection and processing sector, for example, clear benefits are generated by the integration of services along the production chain, given the possibility of offering guarantees on the volume and quality of service demand and, in turn, attracting public and private sector interest towards creating certainty on plant planning. In contexts with less social, political or entrepreneurial cohesion, where such integration has not been possible, situations of uncertainty have become chronic that

---

[9] Acronym for "Not In My Back Yard".

have at times led to progressive management atomisation along the production chain (e.g., in the collection and transport phases) and to the consequent loss of capacity to attract investment.

In this sense, the credibility of the central strategic framework, of plant programming on a regional level and of firms that can compete for the local market tend to mutually condition each other.

Conversely, it is intuitively clear that the pursuit of economies of scale can correspond to a trade-off with respect to the diversification capacity of technological chains. In the case of waste, the transition towards "zero-landfill" that should lead to increasing opportunities for small niche players (see the recovery chains of different forms of materials) would in this sense find in the presence of plant operators a strong entry barrier reducing the variety of conceivable scenarios.

Depending on the objectives pursued, the policy makers' choices must be weighted on a wide range of possible behaviours by all actors along the production chain. In the example mentioned above, it could well be that a material recovery target, when not intrinsically based on positive economies, may find its solution not through the proliferation of competing activity on a local level (in the case of free-riding), or through promoting dominant infrastructures at regional level (in the case of supply chain economies focused on energy conversion), but rather through the creation of consortia on a central level supported by demand guaranteed by mandatory targets.[10]

In more general terms, the balance of central and peripheral administrative components can prevent management atomisation at the local level and, with this, the tendency of municipalities to shield against the effects on the user cost of monopolistic management and to clamp down on investments through the crystallisation of the pool of customers.

In Italy, the importance of such a balance between local and central programming levels has been understood since the time of the earliest designs in the introduction of "optimal territorial ambit authorities"[11] (AATO), theoretically authorities called on to mediate the search for supra-municipal economies of scale with focalisation strategies enabled by operating in environments with limited heterogeneity.

In this regard, for urban water and sanitary services, authorisation acts[12] were issued based on the logic of negotiating rates resting on the analysis of the operator's economic and financial plan. As a point of weakness, this has meant that following privatisation the risks of possible information asymmetries that could still favour the positions of monopolistic rent have gone from strength to strength.

The correct scaling of the system is therefore difficult to establish on paper and more so to implement in a single solution in practice.

And again, looking at the historical development of environmental services, further criticalities emerge from the fact that it is difficult to find complete autonomy

---

[10] This, for example, is the case of waste paper collection, an industry where minimum targets for recycling are set, and free activities of operators have been promoted and sustained by the proceeds of an environmental contribution shown on the invoice on all packing released for consumption.

[11] Through the 1994 Galli Law for water services and the 1997 Ronchi Decree for environmental health services.

[12] Produced by municipalities, provinces or competent homogenous territorial authorities.

between the AATOs and operators. For the same reasons, it is currently not rare to come across difficulties in negotiation processes between public and private actors due to the implications associated with the transfer of ownership of plants[13] or difficulties in finding external private partners who are able, aside from taking over the plant, to bring with them the appropriate technical skills.[14]

Having rejected the hypothesis at national level to proceed by adopting the Anglo-Saxon model,[15] where the individual phases of the production chain are put out to tender, and instead preferring to maintain management unity and, with this, the theoretically greater controllability of service and tariff formation for citizens, the risks of excessive rigidity of the chain can only be managed in an evolutionary perspective through successive management negotiation stages.

Under a theoretical profile – not only immediate for the correct management of tenders but also prospective in order to support effective directive and planning actions – a link between the different levels of governance of the system could thus be effectively guaranteed by a national authority endowed with the responsibility of monitoring and comparing the individual experiences in place. In practice, some steps have been taken in this sense with the establishment of the Antitrust Authority "Autorità Garante della Concorrenza e del Mercato", a central structure that judges devolvements in-house, and the creation of the "Committee for the Control and Use of Water Resources". However, as reported by the same committee, many steps remain to be taken on a number of critical issues in the management of devolvements by AATOs. In turn, unresolved criticalities are ascribable to the committee's lack of operational skills and competencies.[16]

Despite the territorial nature of environmental services, the responsibility of monitoring, knowledge formation and the application of benchmark mechanisms that can ensure the adequate protection of local communities is again placed at the central level.

In this perspective, given the scarcity of verified data available, the correct identification of metrics to measure performance is today the subject of methodological debate. These metrics should hopefully be based on direct quality parameters (e.g., input–output balances, performance management, complaints management time, etc.). Despite this, in the absence of direct intervention by legislation, these are currently more frequently superseded by indirect parameters (e.g., limits imposed by regulation, capital invested, certification and management procedures, etc.). Also for this reason, given the need for common methodological references, the voluntary regulation has been activated, arriving, as in the case of water through the International Organization for Standardization and the International Water Association, at the introduction of a first key series of international quality standards.

---

[13] Consider, for example, the constraints on the stability agreements of the municipalities involved.

[14] The alignment of timings in the privatisation paths throughout the country has led to the opportunity for key incoming investors to select the most interesting opportunities, thus leaving a large number of areas at risk of an absence of necessary skills.

[15] Bennet and Iossa (2006), analysing possible management schemes for public service infrastructures, observe the dependence of the solution on the specificity of the assets and the risk of demand fluctuations.

[16] In particular, signalling the need for interventions to overcome lack of knowledge on the conditions of the network plant, a prime factor that can disadvantage market competition.

The need for the integration of functional units and the coexistence of evaluation objectives at the local, regional and global scale suggest the boundaries for the analysis of these service models in the development of the Life Cycle Assessment and Life Cycle Costing approaches.

10.3
Integrated cycles in environmental services (waste, water) and life cycle management

As mentioned above, the integrated cycle model in environmental services has different variations depending on the application of the context in which it operates. The differences, which can be traced primarily to the extension of management, the network of actors operating in it and their way of interacting in plant programming phases, find justification in the tendency of socio-economic and different cultural contexts to adapt – each developing their own interpretations – to a shared regulatory framework that is constantly evolving.

A great deal of management variety can be observed where performance is often determined by the thoroughness with which the transition from a public service regime to a market system regime has matured, namely the extent to which the responsibility of the service has extended to the consumer, and how much this has remained anchored in the public sector, the extent to which this has resulted in a redistribution of the system costs, the extent to which it has moved from the service provision obligation to the implementation of a negotiated system of rules, and the extent to which the public sector has assumed an independent role in the authorization, control and suppression of inappropriate behaviour.

The essence of the integration of the cycle in environmental services is also manifested in the way in which decisions on any link in the chain have an impact on the entire system and, therefore, the entire life cycle of the service.[17]

These are, for example, the consequences that certain interpretations of the unity of management services for the collection and disposal of municipal waste have had in their effective capacity to reduce the use of landfills: a comparison of municipal waste management plans with hazardous waste management plans shows that the occlusion of communication between the two worlds has effectively precluded the possibility of increasing the number of operators of the separation and recycling activities.

The limitations of service life cycle planning at times derive from the statutory provisions that require operations management to be resolved locally. The advantages of better segregation of responsibilities and the more direct verifiability of management can be counterbalanced by the greater rigidity of service costs in the chain. For example, the regulatory limits for waste transportation – justified on the

---

[17] See the analysis by Eriksson et al. (2005).

system level by the need to prevent dumping[18] – must be evaluated carefully by observing on the one hand the economies of densities typically found in the collection phase and, on the other hand, the economies of scale for the disposal phase. An incorrect definition of the optimal territorial ambit (e.g., due to the segregation of less favourable areas), taking into account the lower appeal that the collection phases can have on potential investors with respect to the disposal phases, impact significantly on the management of the entire service chain.

In some cases, even a *de facto* monopolistic situation in the early stages of disposal, reducing the interest of investors in the remaining stages of the supply chain, could, in the short run, render the industrial cost chain too inflexible and, in the long run, introduce significant barriers in the cost-opportunity of the evolution of the technological paradigm adopted.

Distortions could also occur if the diseconomies of the system and the difficulties in offsetting costs within the various stages lead towards the creation of multi-utilities able to operate at all stages from collection to disposal with the focus on performance moving from individual parts of the chain to its overall design.

The planning of integrated cycles of environmental services, seeking in the first instance to ensure the achievement of an acceptable level of service for the universality of the population, can introduce barriers to the organisation of forms of collection and limit free competition between the forms of disposal that are determined by the theoretical availability of alternative processes made available today by widespread attention to different forms of recovery of different product categories.[19]

In an environmental services market, which to date has tended to be characterised by in-house or incumbent management, or rather, by formerly monopolistic firms in recently liberalised markets, the operators have tended to draw their benefits from a position of large initial advantage (Biondi and Frey 1998). However, we must not forget that the same technical and economic–financial conditions that impede the succession of competitors also constitute a constraint to the entry of potential partners should they be needed. In a scenario where firms adapt their business model only to a limited extent to position themselves in segments of the value chain of higher value added, both in the waste and water sectors, a problem of a lack of mobility of competencies has been created. This scarcity of intellectual resources is today amplified, against the same reform logic, by competition induced by the simultaneous activation of privatisation procedures in the country. Unless the situation is quickly resolved, it could take years to set new corrective scenarios in motion.It is under the pressure of this certainty that, in setting targets for long-term evolution, many players in the system identify in the constitution of mixed and solid public–private entities, capable of mediating the trade-offs so far analysed, the best compromise to rapidly launch the reform system.

---

[18] Originally coined to describe a procedure for transferring a service on a foreign market at a price that is lower than the sales price in the market of origin, today the concept is extended to the transfer of environmental treatment processes in markets with lower social resistance (lower sensitivity or greater organisational capacity) towards the connected infrastructurisation.

[19] Lombrano (2009), with reference to the "Italian case", points out that in the desired transition to high-intensity recovery sectors, a limiting factor may be the smaller size of the collection areas.

According to many of these operators, the best practices that are taken as reference for setting their own industrial strategies also offer unique evidence of the priorities to be addressed to improve the environment in which they operate. This is the case, for example, of the Lombard A2A or the French Veolia, major international multi-utilities counting amongst their keys for success the capacity to regulate the service according to technical–economic rationality criteria, controlling and enforcing rules dictated by experience and knowledge gained, and the capacity to attract and develop the best professional skills on the market.

It is therefore not surprising that regulation efficiency and investment in knowledge are amongst those priorities that operators most frequently bring to the attention of regional and national policy makers to initiate an enduring innovation process system.

When considering water and material life cycles as well as the fundamental objectives of the environmental services sector, the impact that the end user has on the potential effectiveness of innovation processes cannot be overlooked.

More and more, in fact, end users perceive these goods (water and consumer products) as commodities and are thus more careful to limit prices or poor service rather than valorise their distinctive characteristics. This leads on one hand to a loss of perception of the differentiation of the service and, on the other, to their role in determining overall system performance. In order to correct this distortion, community policy makers have recently promoted measures to establish the principle of extended responsibility of the producer, measures in which the forms of application also include raising awareness and market correction by displaying the environmental contributions on prices of goods purchased.

Looking at the future of integrated cycles of environmental services, in order to pursue effectiveness and efficiency in the resource life cycle management, the concept of "final" consumers must be overcome. Consumers have to be fully and actively placed at the centre of a chain of services, whose boundaries must be dictated by knowledge, skills and resources to manage constantly evolving economic and environmental balances.

10.4
Integrated energy cycle and management

Among the public network services analysed here, the distribution of energy (in the form of electricity, gas or heat) is certainly the most dynamic and vibrant in terms of attraction of investment capital.

In fact, having triggered the privatisation process on traditionally large oligopolistic state companies, the effects of uncertainty of the regulatory framework[20] and the

---

[20] Legislature was able to provide a detailed framework for information and capillary knowledge on the functioning of the entire system.

tendency for local governments to impose widespread social tariffs to protect their citizens have been felt less in this sector.

Unlike environmental services, in Italy the energy supply chain seems to be supported to a lesser extent by the presence of national endogenous energy resources. The nuclear option was rejected in the 1987 referendum and Italy is without significant reserves of fossil resources. In its dependence on imports from abroad[21] and significant use of natural gas[22] the national energy system has two significant anomalies compared to the community scenario.

It is thus no coincidence, despite the entrepreneurial maturity of the sector – unlike those analysed above – that domestic and industrial use of energy services are more expensive than the EU average.[23]

The difficulty of attracting the best skills is here replaced, at the enterprise level, by the high palatability of employing young graduates and, on a country system level, by the great investment interest shown by foreign companies, largely attracted by the significant incentives to producers of energy from renewable sources. In these circumstances, the risk of becoming a destination for business colonialism stimulates domestic firms to confront the international competitive landscape.

The evolution of energy companies, with the gradual consolidation of significant investments, has thus indicated the multi-utilities model as winning. The development of new knowledge and the related capacity to generate value at low cost through better utilization of available resources and more efficient use of the firm's "core competencies" have led to increasing potential for market diversification. Economies of scope have thus been triggered between primary and accessory services, such as the design, financing, construction and maintenance of distribution plants, the supply of energy of certificated origin, trading in allowances of CO_2 emissions, the management of district heating networks and provision of energy management services.

Similarly, much effort has been expended in the search for innovative business models for the service component. The resulting initiatives range from partnerships with other operators (see telecommunications, motorways, insurance sectors, etc.) to stock market positions diversified by business units.

It is precisely the polyhedric nature of the service offer that has today inspired the energy industry's new interest in strengthening ties with the territory. In the past, service uniformity meant that these ties were not a determinant of strategic planning. Indeed, this has prevented the morphological heterogeneity of the offer that is unfavourable for strengthening competitiveness in core activities and, consequently, rewarding dominant investments. Today, the valorisation of local resources is instead placed at the centre of complementary offers, no longer considered an element of distraction but an integral part of the offer.

---

[21] The value of net imports/primary energy consumption in 2009 amounted to 87.7% in Italy compared to an EU-27 average of 56.5% (Source: Enerdata).

[22] The value of primary energy generated from natural gas in 2009 in Italy was 38.1% compared to an EU-27 average of 25% (Source: Enerdata).

[23] In the second half of 2007, for Italian residential customers with annual consumption between 2500 and 3500 kWh, an average cost of 22.95 euros/kWh was recorded against the European average of 14.20 euros/kWh; the final price before tax for Italian industrial consumers was instead 9.42 euros/kWh against a European average of 8.86 euros/kWh (Source: AEEG).

A similar dynamic is also particularly interesting for the future development of re-newable energy. In this segment of the market, which due to its size and dynamism has not yet seen the volumes of investments that can crystallise the prevalence of one technology over another, the elements that guarantee the diversity of the pos-sible evolutionary trajectories of their services are the widespread distribution of heterogeneous resources throughout the territory and, depending on the context of use, the capacity to attract innovative technologies.

Opening the market to competition has also meant that the service component has been developed, on one hand, as a means to attract the customer and increase his loyalty and, on the other hand, as a driver of chain efficiency.

Under the energy life cycle management perspective, investments in widespread connectivity and active monitoring of network devices are to be considered, amongst others, the fundamental anchors that promote the realisation of smart grids that would allow the integrated management of on-demand service.

In addition to meeting the new demands of increased and distributed production driven by the spread of private plants using renewable sources, this development would allow producers and consumers to interact, to determine in advance the de-mands of consumption and to adapt electricity production and consumption accord-ingly.

Thus, new prospects would open up for the integrated management of the cycle, alongside greater efficiency of the energy system,[24] enabling consumers to benefit from new services such as new forms of pricing, management of exchanges in higher value-added networks, spread of electric and hydrogen vehicles, etc.[25]

Such inter-sector contamination in the future could become another important en-abler for the creation of business areas favouring system integration (e.g., developers of user–network interfaces) that, in turn, could act as catalysts for the diversification processes of technology providers in those supply chains that intersect with the en-ergy chain (e.g. suppliers of electric engines for vehicles).

The potential explosion of activities of multi-utilities in the transport[26] and telecommunications[27] sectors today foresees the need for new international regula-tions and the creation of national supervisory authorities able to support a harmonic evolution of a new supply chain paradigm only a few years after the widespread privatisation of energy services.

---

[24] Better management of supply and demand would lead to a reduction of energy losses due to technical failures and fraud, the possibility to program consumption in different time phases and, ultimately, the reduction of CO_2 emissions.

[25] The recent interest in the energy sector of ICT giants, such as Google and Apple, demonstrate the attractiveness of the business models that this integration information technology would generate.

[26] See for example Enel Distribuzione's Electric Car project.

[27] See for example so-called "Smart Home Energy Management".

10.5
Conclusions and policy implications

The environmental and energy services sector is going through a period of extensive transformation.

Albeit with significant specificities for each sphere, vertical integration and user expansion represent the challenges introduced by market liberalisation for the majority of operators.

With regard to waste as well as energy services, policy makers, which have assumed the role of guarantors for the provision of minimum services to citizens (sometimes also through the maintenance of a public component of management), are able to directly and indirectly influence diversification and competition between players, especially those oriented to integrating additional services along the supply chain that can potentially affect the life cycle of the resources in question (water, materials consumption or energy).

Recently, rules have been issued around a particular class of players in the waste and energy services sectors: consumers.

These subjects are increasingly considered – with evidence to hand – as able to play an active role in determining the degree of success or failure of the proposed management models. The possibility of affecting the life cycle of services through actions on customers is common among different experiences in increasing consumers' sensibility over the concept of an "extended responsibility".

In general terms, the differences in market determinants and maturity of experiences do not make it possible to transfer policy approaches learned in one domain to another. Consequently, no common solutions can be identified on paper to optimally address the certain, and by now proximate, evolutionary challenges of integrated environmental and energy cycles.[28]

Nevertheless, the evidence analysed suggests the possibility of being able to generalise in the presence of four critical requirements typically under the jurisdiction of policy makers:

1) control of the programming functions;
2) the effectiveness of regulation;
3) equity in the introduction of forms of support;
4) the protection of free competition.

On these points, in fact, rests the ability to attract or repeal the necessary volume of investments.

Commonly found among the most frequent programming failures are, for example, the uncertainties and diseconomies of integration endured by new entrants who have to face the market in the absence of reliable agreements with the various service operators in the chain. Even more significant are the uncertainties related to the implementation of plant development initiatives. In all these sectors, in fact, the stress

---

[28] Bel and Warner (2008), analysing performance recorded in a panel of privatisation cases in the water and waste sectors, did not identify the generalisability of optimal choices.

on performance is at least equal to the problems related to the creation of social consensus on individual initiatives and the resolution of the related debate on the most sustainable way to access use of resources at the local level. Policy makers are thus called on to act as mediators between these conflicting needs.

With reference to the regulatory system, it can frequently be observed how this contributes to determining the extent of possibilities for private investors through, for example, the use of legal monopoly or market competition. On these aspects, the presence of independent national authorities is considered to be able to potentially mitigate the criticalities, present in every context analysed, related to the capacity to ensure equity in access to the market through the single application, however sophisticated, of formal public selection instruments.

The green economy is recognised as having increased planning of systemic efficiency in the use of environmental resources – the real key to optimising costs along the life cycle of environmental and energy services. Besides that, future initiatives of private–public partnerships[29] could acquire greater importance, especially in the case of so-called "hot works". Similarly, the emergence of new business entrants is expected, especially in the case of markets supported by environmental subsidies.

In any case, investments in knowledge and continuous learning of the rules of the system have decisive impact on future development scenarios. If, for example, competition in the market for waste and energy services now appears to be open only to residual capacity (e.g., niche activities implanted in traditional chains), the progressive servitisation of these sectors should spontaneously contribute to shaping the relationship between distinctive and complementary activities. Understanding the functioning of the supply chain and the dynamics of its value chain will therefore mean facilitating the transition to market rules for the introduction of innovations in the integrated service cycle.

Following a widely adopted principle in environmental policies, according to which a system, to ensure the sustainability of its design, must be consistent with global principles and be able to redistribute the generated values to the actors who implemented it, the organisational solutions should spontaneously be affirmed based on the needs of the individual contexts. In line with this, policy makers should not so much pursue the establishment of unitary models as much as the identification of instruments to verify their effectiveness in fulfilling the overall objectives.

References

Bel G, Warner M (2008) Does privatization of solid waste and water services reduce costs? A review of empirical studies. Resourc Conserv Recycling 52:1337–1348

Bennet J, Iossa E (2006) Building and managing facilities for public services. J Public Econ 90:2143–2160

Biondi V, Frey M (1998) Le società di ingegneria nel settore ambientale. In: Genco P, Maraschini F (eds) L'Ingegneria impiantistica. Il Mulino, Bologna

---

[29] Prasad (2006), analysing the distribution and water treatment sector, identifies the drive for the public–private partnership proposal in the ease with which failures are generated in the privatisation processes.

Bishop P (2008) Spatial spillovers and the growth of knowledge intensive services. J Econ Social Geogr 99:281–292

Brusco S, Pertossi P, Cottica A (1995) Mercato, cattura del regolatore e cattura del controllo. Economia e Politica Industriale 88

Eriksson O, Carlsson M, Frostll B et al (2005) Municipal solid waste management from a systems perspective. J Cleaner Prod 13:241–252

ICLEI (1994) Charter of European cities and towns towards sustainability. The Aalborg charter. ICLEI European Secretariat, Freiburg

Liebowitz SJ, Margolis SE (1995) Path dependence, lock-in and history. J Law Econ Organ 11:205–226

Lombrano A (2009) Cost efficiency in the management of solid urban waste. Resourc Conserv and Recycl 53:601–611

Miles I (2006) Innovation in services. In: Fagerberg G, Mowery DC, Nelson RR (eds) The Oxford handbook of innovation. Oxford University Press, Oxford

Picazo-Tadeo A, Saez-Fernandez F, Gonzalez-Gomez F (2008) Does service quality matter in measuring the performance of water utilities? Utilities Policy 16:30–38

Prasad N (2006) Privatisation results: private sector participation in water services after 15 years. Dev Policy Rev 24:669–692

Symbola (2010) Green Italy. Un'idea di futuro per affrontare la crisi. Quaderni di Symbola

UNESCO (2001) Dichiarazione universale dell'UNESCO sulla diversità culturale. Sommet mondial sur le développement durable, Johannesburg

Warner M, Bel G (2008) Competition or monopoly? Comparing privatization of local public services in the US and Spain. Public Admin 86:723–735

WCED (1987) Our common future. Oxford University Press, Oxford

INNOVATION LAB
Smart cities: the cities of the future

MAINS Master, academic year 2009/2010
People and companies involved in the InnoLab:
Students: Maddalena Caracciolo, Francesco Costanzo, Andrea Paraboschi
and Matteo Pastore
Companies: Ericsson Telecomunicazioni, IBM Italia, Intesa San Paolo and
Vodafone Italia
Professors: Lino Cinquini and Riccardo Giannetti

1. The problem
The basic question that framed this project is the increasing complexity of
urban management. Exploring the causes of this complexity, at least two
key factors can be identified: on one hand, the increasing size and crowding
of cities, and secondly, the lack of instruments to support the decisions of
city managers.

The latest UNO and INED projections estimate that world population
will increase from 6.8 billion (2010) to 9.15 billion by 2050 and that the
percentage share of residents in urban areas will rise from 50% to 75%.
It is thus clear that the focus of municipalities is moving towards finding
smart, high-tech and innovative solutions that help adapt the service offer
to increasingly growing and exacting demand from citizens and businesses,
and at the same time, ensure high levels of employment, productivity and
social cohesion.

This increasing complexity makes the job of city managers job even
more complicated. Faced with a fixed budget for the innovation and renewal of services offered to the community, they must decide (without an
adequate panel of decisional instruments) which interventions to prioritise
over others. The decision-making process is further complicated by the heterogeneous characteristics and the different types of solutions that can be
implemented, as well as by the fact that not all services can be considered
smart irrespective of the context of reference.

The work undertaken by the team was also framed in a wider context that
refers to the broader directives issued recently: the European Commission
through the European 20–20 directive, presents medium-term objectives on
the economic, social and environmental front to ensure that growth is simultaneously smart, inclusive and sustainable.

The necessity and demand to respond to community needs, the numerous deployable smart services, a limited budget and an urban fabric that
requires rapid action are key elements from which a model was developed
that could support city managers in the decision-making process and prioritise interventions while maximising cost-effectiveness.

2. Work methodology

The analytical model created by the team is based on a number of regulations expressed by the main international government agencies on sustainability. The first step was to classify sustainability into three main categories: environmental sustainability, social sustainability and economic sustainability.

These three strongly linked categories are the three key aspects of a city's growth and fulfilment. Each service offered to citizens by the urban context under analysis was thus evaluated on the basis of these three pillars of reference.

To do this, the model, through a series of interactions that take into account a large number of variables related to the urban context, assigned a quantitative assessment (expressed in percentage terms) to each category of sustainability for each service.

Through subsequent interaction that also considered the development plan of the municipal administration, a unique score was obtained for each service. This univocal and synthetic judgement can be used to compare the level of sustainability of different services. Through this comparison, a type of classification can be achieved that allows city managers to understand where the greater priorities lie.

The hypothesised model was successfully applied to the Municipality of Pisa. Through continuous dialogue with municipal officials, a model was tested on a sample of six services comprising both completely innovative solutions and services that bring incremental innovations as well as those currently offered by the city.

The working group performed this task in a top-down fashion, decomposing the initial problem into the following sub-problems: the definition of a smart service portfolio to be implemented, understanding and analysing the benefits arising from the implementation of each service, the design of a synthetic model to calculate the overall benefit attainable globally, the comparison of different services and the validation of the model.

In the first phase, the team focused on scouting smart projects implemented in urban domestic and international contexts. A further effort was made in the search for services that do not yet exist but are achievable in the near future, starting from today's technological environment.

The analysis considered 15 Italian plus 20 foreign players, for a total of over 40 projects examined, and led to the definition of a portfolio of 52 possible smart services to implement.

Continuing the process of analysing the project context required understanding the benefits of the services and the particular problems and needs of end users, namely public administrations. While the first part of the process consisted in continued scouting, the second was addressed via numerous meetings held with officials of the municipal administration of Pisa. The process output consisted in mapping the expected benefits and the definition of the relevant performance indicators.

3. Proposed solution

Once a comprehensive overview of the needs of public administrations had been obtained as well as the possible solutions available to date, the team proceeded with the design of a synthetic decision support model. In this phase, the approach consisted in a decomposition of the initial problem. The working group defined and developed the following activities: identification of the cause–effect relationship between the expected benefits and relative indicators, identification of a set of core indicators for the synthetic evaluation of the benefits expected for each category of service and identification of a bottom-up method in order to express a qualitative judgment of the overall service.

Once the design activities were completed, the working group applied the assessment model to a subset of services of particular interest to the council of Pisa in order to test its validity.